An Introduction to
Singular
Integrals

An Introduction to
Singular
Integrals

Jacques Peyrière
Université Paris-Sud
Orsay, France

高等教育出版社·北京
HIGHER EDUCATION PRESS   BEIJING

Society for Industrial and Applied Mathematics
Philadelphia

10 9 8 7 6 5 4 3 2 1

Publications Director	Kivmars H. Bowling
Acquisitions Editor	Paula Callaghan
Developmental Editor	Gina Rinelli Harris
Managing Editor	Kelly Thomas
Production Editor	Louis R. Primus
Copy Editor	Samar Nour-El-Deen
Production Manager	Donna Witzleben
Production Coordinator	Cally A. Shrader
Graphic Designer	Doug Smock

Library of Congress Cataloging-in-Publication Data
CIP data is available at *www.siam.org/books/ot159*

Contents

Preface

The topics developed here have been standard knowledge for at least forty years. Our aim is only to provide newcomers with a basic knowledge of some tools in real analysis: the Hardy–Littlewood maximal operator, Calderón–Zygmund theory, Riesz transforms, Littlewood–Paley theory, Fourier multipliers, H^1 and BMO spaces, and interpolation of operators (real and complex methods). So, it would be in vain to search this book for the present trends of research in this field. We focus on singular convolution operators and develop the Calderón–Zygmund theory in this framework, although more general kernels are mentioned in passing. All this material is of common use in various parts of analysis: harmonic analysis, partial differential equations, signal processing, etc. But this book does not provide such developments in order to respect reader's motivations. This book should be thought as a mere introduction to the study of more complex monographs.

We seek concision, but this text is self-sufficient: there is no assertion without a proof, although some readers may find some of them a bit laconic. This means that some work is expected from the reader: he should read this book with pen in hand. At the end of each chapter a few exercises are proposed with sufficient hints so that a careful reader can solve them himself. Some exercises are just tests for comprehension, while others are important complements, the inclusion of which in the main text would have blurred the exposition.

The prerequisites are a basic knowledge of functional analysis, some acquaintance with measure and integration theory, and a certain familiarity with the Fourier transform in Euclidean spaces. Nevertheless some background material is provided in the appendices. This mainly deals with the use of distribution functions, which, of course, probabilists know very well, but which is not common knowledge among undergraduate students in mathematics.

A few subjects have been selected in this teeming field. As said above, we tried to go directly to the essentials. This is why some proofs, designed for this purpose, may not be straightforwardly transposable to a broader context. In the same spirit we lazily defined the H^1 space by means of atoms. This is not the initial definition, but it is very convenient to establish its duality with BMO. We preferred to develop the Littlewood–Paley theory by using vector singular integrals rather than all the apparatuses of systems of conjugate harmonic functions. Our approach is shorter but less powerful.

A chapter is devoted to Fourier series. Another chapter deals with singular integrals on other groups so that, in particular, it allows one to handle $\mathbb{Q}_p$, the field of p-adic numbers, as well as $\mathbb{F}_q((1/x))$, the field of Laurent series on a finite field.

The last chapter deals with examples of interpolation theorems.

This book grew out of lectures held at Université Paris-Sud at Orsay and at Tsinghua University at Beijing.

I warmly thank Professor Yao Jia-Yan, from Tsinghua University, not only for his careful reading of the manuscript and a wealth of valuable suggestions, but also for undertaking the translation of this book into Chinese. I also extend my thanks to Professor Arnaud Durand, from Université Paris-Sud, who used a preliminary version of this book as a textbook while teaching at Tsinghua University. His remarks contributed to improving the manuscript. Last, I would like to express my gratitude to Professor Wen Zhi-Ying, from Tsinghua University, who gave me the opportunity to stay and teach in prestigious Chinese universities, and to conduct research with many Chinese colleagues.

Chapter 1

The Hardy–Littlewood maximal operator

1.1 ▪ The Hardy–Littlewood operator

Definition 1.1. *If f is a locally integrable function on $\mathbb{R}^n$, one sets*

$$\mathsf{M}f(x) = \sup_{r>0} \frac{1}{|\mathsf{B}(x,r)|} \int_{\mathsf{B}(x,r)} |f(y)|\, \mathrm{d}y. \tag{1.1}$$

$\mathsf{M}f$ *is called the* Hardy–Littlewood maximal function *and the sublinear operation* $f \longmapsto \mathsf{M}f$ *the* Hardy–Littlewood maximal operator.

Theorem 1.2.

1. *There exists a constant $C > 0$ (depending on the dimension n) such that, for all f and all $t > 0$, one has*

$$|\{\mathsf{M}f > t\}| \le \frac{C}{t} \|f\|_1. \tag{1.2}$$

2. *For any $p \in (1, +\infty]$ there exists $C_p > 0$ (depending on n) such that, for any f, one has*

$$\|\mathsf{M}f\|_p \le C_p \|f\|_p. \tag{1.3}$$

Comments

In Definition 1.1, one may only consider rational r. Also it does not matter whether balls are closed or open.

Let χ_r stand for the normalized indicator function of $\mathsf{B}(0,r)$:

$$\chi_r(x) = \frac{1}{|\mathsf{B}(0,r)|} \mathbf{1}_{\mathsf{B}(0,r)}. \tag{1.4}$$

Then

$$\frac{1}{|\mathsf{B}(x,r)|} \int_{\mathsf{B}(x,r)} |f(y)|\, \mathrm{d}y = \frac{1}{|\mathsf{B}(0,r)|} \int_{\mathsf{B}(0,r)} |f(x-y)|\, \mathrm{d}y = |f| * \chi_r(x).$$

Therefore

$$Mf = \sup_{r>0} |f| * \chi_r.$$

But, due to properties of the convolution operation, the functions $|f| * \chi_r$ are continuous. So, as the supremum can be taken over rational r, Mf is measurable. In fact, one can say more: Mf, being a supremum of continuous functions, is lower semicontinuous.

Proof of assertion 1.

We need the following lemma.

Lemma 1.3. *Let $\mathscr{B}$ be a finite collection of open balls in $\mathbb{R}^n$. We can find in $\mathscr{B}$ disjoint balls $B_1, B_2, \dots, B_k$ such that, for every $B \in \mathscr{B}$, we have*

$$B \subset \bigcup_{1 \le j \le k} B_j^*,$$

where B^* stands for the ball with the same center as B and whose radius is three times that of B.

Proof. Set $\mathscr{B}_0 = \mathscr{B}$. Let B_0 be one of the elements of $\mathscr{B}_0$ of maximum radius, and set $\mathscr{B}_1 = \{B \in \mathscr{B}_0 \ : \ B \cap B_0 = \emptyset\}$. If $\mathscr{B}_1$ is nonempty, choose for B_1 one of the elements of $\mathscr{B}_1$ of maximum radius, and set $\mathscr{B}_2 = \{B \in \mathscr{B}_1 \ : \ B \cap B_1 = \emptyset\}$. And so on, until reaching an empty collection of balls. Take $B \in \mathscr{B}$. If B is one of the selected balls $B_1, B_2, \dots$, there is nothing to prove. If it is not, there exists $j \ge 0$ such that $B \in \mathscr{B}_j \setminus \mathscr{B}_{j+1}$. This means that $B \cap B_j \ne \emptyset$ and that the radius of B is less than or equal to the one of B_j. It results that B is contained in B_j^* (draw a picture).   □

Now we can proceed to prove the first assertion of the theorem. Consider a compact subset K of the set $\{Mf > t\}$. For each $x \in K$ there exists an open ball B_x centered at x such that

$$t < \frac{1}{|B_x|} \int_{B_x} |f(y)| \, dy. \tag{1.5}$$

We extract a finite cover of K from the cover $\{B_x\}_{x \in K}$ and apply Lemma 1.3: there are points $x_1, x_2, \dots, x_k$ in K such that $K \subset \bigcup_{1 \le j \le k} B_{x_j}^*$. Then, by taking into account (1.5) and the fact that the balls B_{x_j} are disjoint, we have

$$|K| \le \sum_{j=1}^{k} \left| B_{x_j}^* \right| = 3^n \sum_{j=1}^{k} |B_{x_j}|$$

$$\le \frac{3^n}{t} \sum_{j=1}^{k} \int_{B_{x_j}} |f(y)| \, dy = \frac{3^n}{t} \int_{\bigcup_{j=1}^{k} B_{x_j}} |f(y)| \, dy$$

$$\le \frac{3^n}{t} \int_{\mathbb{R}^n} |f(y)| \, dy.$$

We conclude by using the inner regularity of the Lebesgue measure:

$$|\{Mf > t\}| = \sup\{|K| \ : \ K \subset \{Mf > t\}, \text{ compact}\}.$$

Exercise 1.5 presents an alternate proof.

Proof of assertion 2.

Let $t > 0$ be fixed for a while. Set

$$g(x) = \begin{cases} f(x) & \text{if } |f(x)| > t/2, \\ 0 & \text{otherwise,} \end{cases} \qquad \text{and} \qquad h = f - g.$$

We then have $Mh \leq t/2$, so

$$\{Mf > t\} \subset \{Mg > t/2\} \cup \{Mh > t/2\} = \{Mg > t/2\},$$

which implies

$$|\{Mf > t\}| \leq \frac{2C}{t} \int_{\{|f| > t/2\}} |f(x)| \, dx. \tag{1.6}$$

This last inequality improves on the first statement of Theorem 1.2 and could be useful in other situations.

By using Lemma A.1, this inequality, and the Fubini theorem, we get

$$\begin{aligned}
\int_{\mathbb{R}^n} Mf(x)^p dx &= p \int_0^{+\infty} t^{p-1} |\{Mf > t\}| \, dt \\
&\leq 2pC \int_0^{+\infty} t^{p-2} \left(\int_{|f| > t/2} |f(x)| \, dx \right) dt \\
&= 2pC \int_{\mathbb{R}^n} |f(x)| \left(\int_0^{2|f(x)|} t^{p-2} dt \right) dx \\
&= \frac{pC2^p}{p-1} \int_{\mathbb{R}^n} |f(x)|^p dx.
\end{aligned}$$

This paragraph above is a particular case of the Marcinkiewicz interpolation theorem (Theorem 8.1).

Exercise 1.7 presents an alternate proof.

1.2 ▪ The Lebesgue derivation theorem

Theorem 1.4. *Let* $f \in L^1_{loc}(\mathbb{R}^n)$. *For almost every* x *in* $\mathbb{R}^n$, *we have*

$$\lim_{r \searrow 0} \frac{1}{|B(x,r)|} \int_{B(x,r)} |f(y) - f(x)| \, dy = 0. \tag{1.7}$$

Proof. Obviously, it is enough to prove this theorem when f is integrable.

Let $\mathscr{K}(\mathbb{R}^n)$ stand for the set of continuous functions with compact support. For any integrable function, say g, set

$$\Omega g(x) = \limsup_{\substack{s \searrow 0 \\ r \leq s}} \frac{1}{|B(x,r)|} \int_{B(x,r)} |g(y) - g(x)| \, dy.$$

Ωg is defined almost everywhere. We have the following facts:

1. $\Omega(g_1 + g_2) \leq \Omega g_1 + \Omega g_2$.

2. $\Omega g \leq \mathsf{M} g + |g|$.

3. $\Omega \varphi = 0$ if $\varphi \in \mathscr{K}(\mathbb{R}^n)$.

Let f be an integrable function. For any $\varphi \in \mathscr{K}(\mathbb{R}^n)$, we have

$$\begin{aligned} \Omega f &= \Omega\big(\varphi + (f - \varphi)\big) \\ &\leq \Omega\varphi + \Omega(f - \varphi) = \Omega(f - \varphi) \\ &\leq \mathsf{M}(f - \varphi) + |f - \varphi|. \end{aligned}$$

Therefore, for any $t > 0$, we have

$$\begin{aligned} |\{\Omega f > t\}| &\leq |\{\mathsf{M}(f - \varphi) > t/2\}| + |\{|f - \varphi| > t/2\}| \\ &\leq \frac{2C}{t} \|f - \varphi\|_1 + \frac{2}{t} \|f - \varphi\|_1 = \frac{2(C+1)}{t} \|f - \varphi\|_1. \end{aligned}$$

This implies that, for all $t > 0$, we have

$$|\{\Omega f > t\}| \leq \frac{2(C+1)}{t} \inf_{\varphi \in \mathscr{K}(\mathbb{R}^n)} \|f - \varphi\|_1.$$

However, as $\mathscr{K}(\mathbb{R}^n)$ is dense in $L^1(\mathbb{R}^n)$, this last quantity is 0. This means that, for every $t > 0$, $|\{\Omega f > t\}| = 0$. By writing

$$\{\Omega f \neq 0\} = \bigcup_{k \geq 1} \left\{\Omega f > \frac{1}{k}\right\},$$

we get $|\{\Omega f \neq 0\}| = 0$.

Definition 1.5. *A point x for which* (1.7) *holds is called a* Lebesgue point *of f.*

Obviously, if x is a Lebesgue point of f, one has

$$f(x) = \lim_{r \searrow 0} \frac{1}{|\mathsf{B}(x,r)|} \int_{\mathsf{B}(x,r)} f(y)\,\mathrm{d}y.$$

Definition 1.6. *If E is a measurable subset of $\mathbb{R}^n$, a Lebesgue point of the indicator function of E is called a* density point *of E.*

In other words, $x \in E$ is a density point of E means

$$\lim_{r \searrow 0} \frac{\mathcal{L}\big(E \cap \mathsf{B}(x,r)\big)}{\mathcal{L}\mathsf{B}(x,r)} = 1.$$

Then almost every point of a set of positive Lebesgue measure is a density point.

1.3 ▪ Regular families

Definition 1.7. *A family $\mathcal{F}$ of measurable subsets of $\mathbb{R}^n$ is said to be regular (for derivation of integrals) if it contains sets of arbitrarily small Lebesgue measure and if there exists $c > 0$ such that, for every $F \in \mathcal{F}$, there exists a ball B centered at 0, containing F, and such that $|F| \geq c|B|$.*

One defines

$$M_{\mathcal{F}} f(x) = \sup_{F \in \mathcal{F}} \frac{1}{|F|} \int_F |f(x - y)| \, dy.$$

If $\mathcal{F}$ is regular, with the notation of the previous definition, one has

$$\frac{1}{|F|} \int_F |f(x - y)| \, dy \leq \frac{1}{|F|} \int_B |f(x - y)| \, dy$$

$$\leq \frac{c^{-1}}{|B|} \int_B |f(x - y)| \, dy \leq c^{-1} M f(x).$$

In other words, $M_{\mathcal{F}} f \leq c^{-1} M f$. Therefore, Theorem 1.2 also holds (with different constants) for the maximal operator $M_{\mathcal{F}}$ instead of M.

Also, a proof following along the same lines as that of Theorem 1.4 shows that

$$\lim_{|F| \to 0, F \in \mathcal{F}} \frac{1}{|F|} \int_F |f(x - y) - f(x)| \, dy = 0 \tag{1.8}$$

for almost every x.

Indeed, it is not difficult to show that (1.8) holds at any Lebesgue point of f:

$$\frac{1}{|F|} \int_F |f(x - y) - f(x)| \, dy \leq \frac{1}{|F|} \int_B |f(x - y) - f(x)| \, dy$$

$$\leq \frac{c^{-1}}{|B|} \int_B |f(x - y) - f(x)| \, dy.$$

Examples of regular families

1. The family of balls containing the origin ($c = 2^{-n}$).

2. The family of cubes containing the origin ($1/c = n^{n/2} v_n$, where v_n is the volume of the unit ball in $\mathbb{R}^n$).

The family of all rectangles with sides parallel to the axes *is not* regular.

1.4 ▪ Control of some convolutions

Lemma 1.8. *Let γ be a nonnegative and nonincreasing function on $(0, +\infty)$. For $x \in \mathbb{R}^n$, set $\psi(x) = \gamma(|x|)$ and suppose that ψ is integrable. Then, for any nonnegative measurable function f, we have*

$$\int_{\mathbb{R}^n} f(x) \, \psi(x) \, dx \leq \|\psi\|_1 M f(0).$$

Proof. First, we consider the case when γ is of the form

$$\gamma = \sum_{j=1}^{k} \alpha_j \mathbf{1}_{(0,r_j)}, \tag{1.9}$$

where $0 < r_j < r_{j+1}$ and $\alpha_j > 0$ for all j. We then have

$$\int_{\mathbb{R}^n} f(x)\,\psi(x)\,\mathrm{d}x = \sum_{j=1}^{k} \alpha_j \int_{B(0,r_j)} f(x)\,\mathrm{d}x$$

$$\leq \left(\sum_{j=1}^{k} \alpha_j |B\,(0,r_j)| \right) \mathsf{M} f(0)$$

$$= \|\psi\|_1 \mathsf{M} f(0).$$

Now, let us consider the general case. For any integer $m > 0$, set

$$\gamma_m = \sum_{j=0}^{m2^m-1} \frac{j}{2^m} \mathbf{1}_{\{j2^{-m} \leq \gamma < (j+1)2^{-m}\}}.$$

Then the sequence of functions $(\gamma_m)_{m \geq 1}$, all of the form (1.9) since γ is nonincreasing, is nondecreasing and pointwise converges towards ψ. We conclude by using the monotone convergence theorem.

A function which depends only on $|x|$ is called *radial*. If $\psi(x) = \gamma(|x|)$, then γ is called the *profile* of ψ. Lemma 1.8 can be reformulated as follows.

Proposition 1.9. *Let φ be a measurable function on $\mathbb{R}^n$ such that the function $\psi(x) = \sup_{|y| \geq |x|} |\varphi(y)|$ is integrable. Then, for $f \in \bigcup_{1 \leq p \leq \infty} L^p$,*

1. $|f * \varphi(x)| \leq \|\psi\|_1 \mathsf{M} f(x)$;

2. $\sup_{t>0} |f * \varphi_t(x)| \leq \|\psi\|_1 \mathsf{M} f(x)$, *where* $\varphi_t(y) = t^{-n}\varphi(y/t)$;

3. *for almost all x,* $\lim_{t \searrow 0} f * \varphi_t(x) = f(x) \int \varphi(y)\,\mathrm{d}y$.

Proof. Fix x and set $h(y) = f(x - y)$. Then we have $\mathsf{M} h(0) = \mathsf{M} f(x)$. So, due to Lemma 1.8,

$$|f * \varphi(x)| \leq \int |h(y)|\,|\varphi(y)|\,\mathrm{d}y \leq \|\psi\|_1 \mathsf{M} h(0) = \|\psi\|_1 \mathsf{M} f(x).$$

This proves the first assertion, from which it is easy to deduce the second.

If $f \in \mathscr{K}(\mathbb{R}^n)$, the limit in the last assertion holds everywhere. Then the proof of the last assertion follows along the same lines as that of Theorem 1.4, but one has to use the following lemma.

Lemma 1.10. *With the same notation as above, if f is continuous with compact support, one has*

$$\lim_{t \searrow 0} \int t^{-n} \varphi(x/t) f(-x) \, \mathrm{d}x = f(0) \int \varphi(x) \, \mathrm{d}x.$$

Proof. See Exercise 1.8

1.5 ▪ Exercises

Exercise 1.1. *A function from a topological space X to $\overline{\mathbb{R}}$ is said to be lower semicontinuous (lsc, for short) if, for all $t \in \mathbb{R}$, the set $\{f > t\}$ is open.*

Let $\{f_i\}_{i \in I}$ be a family of continuous functions from X to $\overline{\mathbb{R}}$. Show that the function $\sup_{i \in I} f_i$ is lsc.

Exercise 1.2. *Let a and b be two numbers such that $a < b$. Compute $\mathsf{M}\mathbf{1}_{[a,b]}$.*

Exercise 1.3. *Let $f = \mathbf{1}_{B(0,1)}$ be the indicator function of the unit ball in $\mathbb{R}^n$. Show that, for $|x| > 1$, $\mathsf{M}f(x) \le \mathsf{v}_n(|x| - 1)^{-n}$. Conclude that, for $p > 1$, $\mathsf{M}f \in L^p(\mathbb{R}^n)$.*

Exercise 1.4. *If $f \in L^1(\mathbb{R}^n)$ such that $f \ne 0$, then $\mathsf{M}f \notin L^1(\mathbb{R}^n)$ (prove that $\mathsf{M}f(x) \ge C/|x|$ for $|x|$ large enough).*

Exercise 1.5.

1. *Let $\rho > 3$. Let $\mathscr{B}$ be a collection of balls of bounded radii. Prove that one can extract from $\mathscr{B}$ a sequence of disjoint balls (B_j) such that*

$$\mathcal{L} \bigcup_{\mathsf{B} \in \mathscr{B}} \mathsf{B} \le \sum_{j \ge 1} \mathcal{L}\mathsf{B}_j^* = \rho^n \sum_{j \ge 1} \mathcal{L}\mathsf{B}_j,$$

   where B_j^* is the ball with the same center as B_j and whose radius is ρ times the radius of B_j.

2. *Give a proof of the first assertion of Theorem 1.2 which does not use the regularity of the Lebesgue measure,*

Exercise 1.6. *Let $f_1, f_2, \ldots, f_m, \ldots$ be a nondecreasing sequence of nonnegative functions in $L^1(\mathbb{R}^n)$. Let f be the pointwise limit of f_m. Show that, for all $x \in \mathbb{R}^n$,*

$$\mathsf{M}f(x) = \lim_{m \to \infty} \mathsf{M}f_m(x).$$

Exercise 1.7 (an alternate proof of the L^p-boundedness of M). *If $f \in L^1_{\mathrm{loc}}(\mathbb{R}^n)$, define*

$$\widetilde{\mathsf{M}}f(x) = \sup \left\{ \frac{1}{|B|} \int_B |f(y)| dy \; : \; B \text{ an open ball, } x \in B \right\}.$$

1. *Show that the set $\{\widetilde{\mathsf{M}}f > t\}$ is open.*

2. *Show that we have*

$$\left|\left\{\widetilde{\mathsf{M}}f > t\right\}\right| \leqslant \frac{3^n}{t} \int_{\widetilde{\mathsf{M}}f>t} |f|\,\mathrm{d}\lambda.$$

3. *Let $p \in (1, +\infty)$. Show that*

$$\int (\widetilde{\mathsf{M}}f)^p\,\mathrm{d}\lambda \leqslant \frac{3^n p}{p-1} \int |f|\,(\widetilde{\mathsf{M}}f)^{p-1}\,\mathrm{d}\lambda \leqslant \frac{3^n p}{p-1} \|f\|_p \left(\int (\widetilde{\mathsf{M}}f)^p\,\mathrm{d}\lambda\right)^{\frac{p-1}{p}}.$$

4. *Let $p \in (1, +\infty)$. Show that $\|\widetilde{\mathsf{M}}f\|_p \leqslant \frac{3^n p}{p-1} \|f\|_p$.*
   Hint: *Use Exercises 1.3 and 1.6 and the preceding inequality.*

5. *Show that*

$$\|\mathsf{M}f\|_p \leqslant \frac{3^n p}{p-1} \|f\|_p.$$

Exercise 1.8. *Let f be a bounded measurable function on $\mathbb{R}^n$, continuous at 0 and such that $f(0) = 0$. Then, if $\varphi \in L^1(\mathbb{R}^n)$, prove that*

$$\lim_{t \searrow 0} \int t^{-n} f(x)\varphi(x/t)\,\mathrm{d}x = 0.$$

Exercise 1.9. *A positive Borel measure μ on $\mathbb{R}^n$ is said to be* doubling *if there exists a constant C such that, for all $x \in \mathbb{R}^n$ and all $r > 0$, one has*

$$\mu\big(\mathsf{B}(x, 2r)\big) \leq C\,\mu\big(\mathsf{B}(x, r)\big).$$

For such a measure, prove that, for all $\gamma > 1$, there exists C_γ such that, for all $x \in \mathbb{R}^n$ and all $r > 0$, one has

$$\mu\big(\mathsf{B}(x, \gamma r)\big) \leq C_\gamma\,\mu\big(\mathsf{B}(x, r)\big).$$

Exercise 1.10. *Prove that, for $\alpha > 0$ the measure (on $\mathbb{R}^n$) $\mathrm{d}\mu(x) = |x|^\alpha \mathrm{d}x$ is doubling.*

Exercise 1.11. *Let μ be a doubling measure on $\mathbb{R}^n$. If f is locally integrable with respect to μ, one sets*

$$\mathsf{M}_\mu f(x) = \sup_{r>0} \frac{1}{\mu\big(\mathsf{B}(x, r)\big)} \int_{\mathsf{B}(x,r)} |f|\,\mathrm{d}\mu.$$

Prove that there exists C such that, for all $f \in L^1(\mu)$ and all $t > 0$, one has

$$\mu(\mathsf{M}_\mu f > t) \leq \frac{C}{t} \|f\|_{L^1(\mu)}.$$

Exercise 1.12. *Let f be an integrable function on $\mathbb{R}^n$.*

Show that if $\displaystyle\int_{\mathbb{R}^n} |f(x)| \log^+(|f(x)|)\, dx$ is finite, then $\mathsf{M}f$ is locally integrable.

Hint: *If B is a ball, write $\displaystyle\int_{\mathsf{B}} \mathsf{M}f \, d\lambda \leq 2\,\lambda(\mathsf{B}) + \int_{\mathsf{M}f > 2} \mathsf{M}f \, d\lambda$, and use Exercise A.8 and the inequality $t\,\lambda(\{\mathsf{M}f > t\}) \leq C \displaystyle\int_{|f| > t/2} |f|\, d\lambda$.*

Exercise 1.13. *For $f \in L^1_{\mathrm{loc}}(\mathbb{R}^2)$, one sets*

$$\mathcal{M} f(x, y) = \sup_{\substack{\varepsilon > 0 \\ \eta > 0}} \frac{1}{4\varepsilon\eta} \iint_{\substack{|t| < \varepsilon \\ |u| < \eta}} |f(x - t, y - u)|\, dt\, du.$$

1. *Let φ be the indicator function of the square $[0, 1] \times [0, 1]$. Prove that $\mathcal{M}\varphi \notin L^1_{\mathrm{w}}(\mathbb{R}^2)$.*

   Hint: *Show that, for $x > 1$ and $y > 1$, one has $\mathcal{M}\varphi(x, y) \geq \frac{1}{xy}$.*

2. *Prove that for any $p > 1$, there exists C_p such that, for all $f \in L^p(\mathbb{R}^2)$, one has*

   $$\|\mathcal{M}f\|_p \leq C_p\|f\|_p.$$

   (Use twice the L^p estimate for the 1-dimensional Hardy–Littlewood maximal operator.)

Exercise 1.14. *Consider a family of rectangles $\{R_t = (-a_t, a_t) \times (-b_t, b_t)\}_{t > 0}$ in $\mathbb{R}^2$ such that, if $0 < t < s$, one has $R_t \subset R_s$.*

1. *Given a finite number n of rectangles $\{\mathsf{S}_j\}_{1 \leq j \leq n}$ of the form $\mathsf{S}_j = (x_j, y_j) + R_{t_j}$, prove that one can find a subset $F \subset \{1, 2, \ldots, n\}$ such that the rectangles $\{\mathsf{S}_j\}_{j \in F}$ are disjoint and*

   $$\bigcup_{1 \leq j \leq n} \mathsf{S}_j \subset \bigcup_{j \in F} \left((x_j, y_j) + 3R_{t_j}\right).$$

2. *If $f \in L^1_{\mathrm{loc}}(\mathbb{R}^2)$, one sets $\mathsf{M}f(x) = \displaystyle\sup_{t > 0} \frac{1}{|R_t|} \int_{R_t} |f(x - y)|\, dy$.*

   Show that if $f \in L^1(\mathbb{R}^2)$, one has

   $$|\{\mathsf{M}f > t\}| \leq \frac{9}{t}\,\|f\|_{L^1(\mathbb{R}^2)}.$$

Chapter 2

Principal values, and some Fourier transforms

2.1 ▪ Operators commuting with translations

If f is a function on $\mathbb{R}^n$ and $a \in \mathbb{R}^n$, $\tau_a f$ is the function $\tau_a f(x) = f(x-a)$. The operators τ_a are called *translations*. They are isometries of all the L^p spaces ($L^p(\mathbb{R}^n)$ is written as L^p for short).

Lemma 2.1. *Let T be a linear operator, bounded on L^p for some $p \in [1, +\infty)$, and commuting with translations. Then if $f \in L^p$ and $g \in L^1$, we have*

$$(Tf) * g = T(f * g).$$

Proof. We have

$$f * g(x) = \int f(x-y)\,g(y)\,\mathrm{d}y = \int \tau_y f(x)\,g(y)\,\mathrm{d}y. \tag{2.1}$$

If $u \in L^p$ and $v \in L^{p'}$, we write $\langle u, v \rangle = \int_{\mathbb{R}^n} u(x)\,v(x)\,\mathrm{d}x$. We also know that there exists a bounded linear operator T' on $L^{p'}$, the adjoint of T, such that $\langle Tu, v \rangle = \langle u, T'v \rangle$.

We have, for all $h \in L^{p'}$,

$$
\begin{aligned}
\langle T(f * g), h \rangle &= \langle f * g, T'h \rangle \\
&= \iint \tau_y f(x)\,g(y)\,T'h(x)\,\mathrm{d}x\,\mathrm{d}y \\
&= \int \langle \tau_y f, T'h \rangle\,g(y)\,\mathrm{d}y \\
&= \int \langle T\tau_y f, h \rangle\,g(y)\,\mathrm{d}y \\
&= \int \langle \tau_y T f, h \rangle\,g(y)\,\mathrm{d}y \\
&= \iint \tau_y T f(x)\,g(y)\,h(x)\,\mathrm{d}x\,\mathrm{d}y \\
&= \langle (Tf) * g, h \rangle.
\end{aligned}
$$

11

We conclude that $T(f * g) = (Tf) * g$.

We can comment on this proof. Equation (2.1) can be written as a vector-valued integral (see Section A.1):

$$f * g = \int \tau_y f \, g(y) \, dy.$$

Then

$$T(f * g) = \int T(\tau_y f) \, g(y) \, dy = \int \tau_y (Tf) \, g(y) \, dy = (Tf) * g.$$

Theorem 2.2. *T is a bounded linear operator on L^2 commuting with translations if and only if there exists a bounded measurable function m such that, for all $f \in L^2$, $\widehat{Tf} = m \, \widehat{f}$. The function m is called the multiplier associated with T.*

Proof. If $m \in L^\infty$ and $f \in L^2$, due to Plancherel identity, $m \, \widehat{f}$ is square integrable, and therefore is the Fourier transform of some $Tf \in L^2$. This defines a bounded operator on L^2. It commutes with translations:

$$\widehat{T\tau_a f}(y) = m(y)\widehat{\tau_a f}(y) = m(y)e^{-2i\pi a \cdot y}\widehat{f}(y) = e^{-2i\pi a \cdot y}\widehat{Tf}(y) = \widehat{\tau_a Tf}(y).$$

Now, consider an operator T bounded on L^2 and commuting with translations.

Take $f \in L^1 \cap L^2$ and $\varphi(x) = e^{-\pi|x|^2}$. Due to Lemma 2.1, we have $T(f * \varphi) = \varphi * Tf = f * T\varphi$. By taking the Fourier transform, we get $\varphi \widehat{Tf} = \widehat{f} \widehat{T\varphi}$, which can be written as $\widehat{Tf} = m \, \widehat{f}$ if we set $m = \widehat{T\varphi}/\varphi$. This means that for all $f \in L^1 \cap L^2$, $\widehat{Tf} = m\widehat{f}$.

Now, we have to show that m is bounded. Suppose that m is not essentially bounded. This means that for any α, which will be chosen later, the set $\{|m| > \alpha\}$ has a nonzero Lebesgue measure. So, it has a density point y_o (see Definition 1.6). This implies that there exists $r > 0$ such that

$$\mathcal{L}\big(\{|m| > \alpha\} \cap B\,(y_0, r)\big) \geq \mathcal{L}\big(B\,(y_0, r)\big)/2.$$

Now, take f such that $\widehat{f}(y) = r^{-n}e^{-\pi|y - y_0|^2 r^{-2}}$. We have

$$\|m\widehat{f}\|_2^2 \geq \frac{e^{-2\pi\alpha^2 v_n}}{2} r^{-n}$$

and

$$\|m\widehat{f}\|_2^2 = \|Tf\|_2^2 \leq \|T\|^2 \|f\|_2^2 = 2^{-n/2} r^{-n} \|T\|^2.$$

So, when α is large enough, we get a contradiction. It results that $\|m\|_\infty < +\infty$. Then it results from the Plancherel identity that, for all $f \in L^2$, $\widehat{Tf} = m\widehat{f}$.

Theorem 2.3. *For a bounded linear operator T on L^1 the following are equivalent:*

1. *T commutes with translations.*

2. *There exists a bounded Borel measure μ such that $Tf = \mu * f$ for all $f \in L^1$.*

Proof. Let $g \in \mathscr{K}$ such that $\int g(x) \, dx = 1$, and set $g_t(x) = t^{-n}g(t^{-1}x)$. Since the functions g_t are bounded in L^1, so are $T(g_t)$. So, there exists a sequence t_j converging to 0 such that the measure $T(g_{t_j})(x)\,\mathrm{d}x$ converges weakly towards a measure μ. More precisely, for all $\varphi \in \mathscr{K}$,

$$\int \varphi \,\mathrm{d}\mu = \lim_{j\to\infty} \int \varphi(x)\,T(g_{t_j})(x)\,\mathrm{d}x,$$

which implies

$$\varphi * \mu = \lim_{j\to\infty} \varphi * T(g_{t_j}) \quad \text{pointwise.}$$

But $\varphi * T(g_{t_j}) = T(\varphi * g_{t_j})$. As T is a bounded operator on L^1, $T(\varphi * g_{t_j})$ converges in L^1 towards $T(\varphi)$. Since L^1-convergence implies the almost everywhere convergence of a subsequence, we get $\varphi * \mu = T(\varphi)$.

We conclude by using the density in L^1 of the continuous functions with compact support.

Lemma 2.4. *Let $p > 1$. If T is a linear operator bounded on L^p and commuting with translations, then, for all q in the interval with endpoints p and p' (in particular for $q = 2$), it is bounded on L^q.*

Proof. If f and g are continuous functions with compact supports, we have $f * T(g) = (Tf) * g$, and these convolutions are continuous. In particular $f * T(g)(0) = (Tf) * g(0)$, which imply

$$\left| \int Tg(x)\,f(-x)\,\mathrm{d}x \right| = \left| \int g(x)\,Tf(-x)\,\mathrm{d}x \right| \le \|g\|_{p'}\|Tf\|_p \le \|T\|_{p\to p}\|g\|_{p'}\|f\|_p.$$

We therefore have $\|Tg\|_{p'} \le \|T\|_{p\to p}\|g\|_{p'}$. Since $\mathscr{S}$ is dense in $L^{p'}$, T defines a bounded linear operator on $L^{p'}$. Also, T has the same norm as operator on L^p or $L^{p'}$.

We can now use the Riesz–Thorin theorem (Theorem 8.6) with endpoints (p, p) and (p', p') to complete the proof.

As seen above, there are characterizations of multipliers for L^1 and L^2. There is no such characterization for L^p, with $p \ne 1, 2$. One of the aims of this book is to provide sufficient conditions for a bounded function to be an L^p multiplier.

2.2 ▪ Principal values

Let K be a locally integrable function on $\mathbb{R}^n \setminus \{0\}$ such that

1. $\sup |x|^n |K(x)| < \infty$;

2. for all r and R such that $0 < r < R$, $\displaystyle\int_{r \le |x| \le R} K(x)\,\mathrm{d}x = 0$.

Let $\varphi \in \mathscr{S}(\mathbb{R}^n)$. For $r > 0$, consider the expression

$$I(r) = \int_{|x|>r} K(x)\,\varphi(x)\,\mathrm{d}x.$$

Due to the second requirement on K, one has

$$I(r) = \int_{r<|x|<R} K(x)\,(\varphi(x) - \varphi(0))\,\mathrm{d}x + \int_{|x|>R} K(x)\,\varphi(x)\,\mathrm{d}x$$

for all $R > r$. Since $K(x)\big(\varphi(x) - \varphi(0)\big)$ is integrable near 0, it results that $I(r)$ has a limit, which will be denoted by $\operatorname{pv} \int K(x)\,\varphi(x)\,\mathrm{d}x$, as r goes to 0. In other words, this defines a tempered distribution $\operatorname{pv} K$: $\langle \operatorname{pv} K, \varphi \rangle = \operatorname{pv} \int K(x)\,\varphi(x)\,\mathrm{d}x$.

There are situations other than that described just above which allow us to define the tempered distribution $\operatorname{pv} K$. See, for instance, Exercise 2.2.

2.3 ▪ Some Fourier transforms

Theorem 2.5. *Let K be a locally integrable function on $\mathbb{R}^n \backslash \{0\}$ such that there exists B so that*

a. $|K(x)| \le B\,|x|^{-n}$;

b. $\displaystyle \sup_{y \ne 0} \int_{|x| \ge 2|y|} |K(x-y) - K(x)|\,\mathrm{d}x \le B$;

c. *for $0 < r < R$, one has $\displaystyle \int_{r \le |x| \le R} K(x)\,\mathrm{d}x = 0$.*

Consider the truncated kernel $K^{(\varepsilon)}$: $K^{(\varepsilon)}(x) = \begin{cases} 0 & \text{if } |x| < \varepsilon, \\ K(x) & \text{if } |x| \ge \varepsilon. \end{cases}$ Then

1. *there exists C depending only on B and n such that the Fourier transform m_ε of $K^{(\varepsilon)}$ has a L^∞ norm less than or equal to C;*

2. *m_ε converges almost everywhere towards a function m which is the Fourier transform of $\operatorname{pv} K$;*

3. *if $f \in L^2(\mathbb{R}^n)$, then $K^{(\varepsilon)} * f \in L^2(\mathbb{R}^n)$ and $\big(K^{(\varepsilon)} * f\big)^{\widehat{}} = \mathrm{m}_\varepsilon\, \widehat{f}$;*

4. *if $f \in L^2(\mathbb{R}^n)$, $K^{(\varepsilon)} * f$ has a limit, Tf, in L^2 as ε goes to 0, and one has $\widehat{Tf} = \mathrm{m}\,\widehat{f}$.*

Proof. Since $K^{(\varepsilon)}$ is a L^2-function, its Fourier transform, say m_ε, is the limit in L^2 as R goes to $+\infty$ of $\mathrm{m}_{\varepsilon,R}(y) = \displaystyle \int_{\varepsilon < |x| \le R} e^{-2i\pi\,x\cdot y} K(x)\,\mathrm{d}x$ (it is also the almost everywhere limit of a sequence $\mathrm{m}_{\varepsilon, R_j}$). We are going to prove that $|\mathrm{m}_\varepsilon|$ is dominated by a bound depending only on n and B.

Fix $y \ne 0$ and consider $R > 2/|y|$. We write

$$\mathrm{m}_{\varepsilon,R}(y) = \int_{\varepsilon < |x| \le 2/|y|} e^{-2i\pi\,x\cdot y} K(x)\,\mathrm{d}x + \int_{2/|y| < |x| \le R} e^{-2i\pi\,x\cdot y} K(x)\,\mathrm{d}x.$$

If $|y| \ge 2/\varepsilon$, the first integral does not exist. Otherwise, we have (due to Hypothesis c.)

$$\left| \int_{\varepsilon < |x| \le 2/|y|} e^{-2i\pi\,x\cdot y} K(x)\,\mathrm{d}x \right| = \left| \int_{\varepsilon < |x| \le 2/|y|} \big(e^{-2i\pi\,x\cdot y} - 1\big)\,K(x)\,\mathrm{d}x \right|$$

$$\le \int_{\varepsilon < |x| \le 2/|y|} 2\,|\sin \pi x \cdot y|\,|K(x)|\,\mathrm{d}x$$

$$\le 2\pi B |y| \int_{\varepsilon < |x| \le 2/|y|} |x|^{1-n}\,\mathrm{d}x \le 4\pi n\, \mathrm{v}_n B.$$

Now, we consider the second integral. It is convenient to use the following notation:

$$\mathsf{C}(r, R) = \{x \in \mathbb{R}^n \;:\; r < |x| \le R\},$$

$$\mathsf{m}_{\varepsilon,R}^{(2)}(y) = \int_{\mathsf{C}(2/|y|,R)} e^{-2\mathrm{i}\pi\, x\cdot y} K(x)\,\mathrm{d}x.$$

Set $z = \dfrac{y}{2|y|^2}$. We have

$$
\begin{aligned}
\mathsf{m}_{\varepsilon,R}^{(2)}(y) &= -\int_{\mathsf{C}(2/|y|,R)} e^{-2\mathrm{i}\pi\,(x+z)\cdot y} K(x)\,\mathrm{d}x \\
&= \frac{1}{2}\int_{\mathsf{C}(2/|y|,R)} e^{-2\mathrm{i}\pi\, x\cdot y} K(x)\,\mathrm{d}x - \frac{1}{2}\int_{\mathsf{C}(2/|y|,R)} e^{-2\mathrm{i}\pi\,(x+z)\cdot y} K(x)\,\mathrm{d}x \\
&= \frac{1}{2}\int_{\mathsf{C}(2/|y|,R)} e^{-2\mathrm{i}\pi\, x\cdot y} K(x)\,\mathrm{d}x - \frac{1}{2}\int_{z+\mathsf{C}(2/|y|,R)} e^{-2\mathrm{i}\pi\, x\cdot y} K(x-z)\,\mathrm{d}x \\
&= \frac{1}{2}\left(\int_{\mathsf{C}(2/|y|,R)} e^{-2\mathrm{i}\pi\, x\cdot y} K(x)\,\mathrm{d}x - \int_{z+\mathsf{C}(2/|y|,R)} e^{-2\mathrm{i}\pi\, x\cdot y} K(x)\,\mathrm{d}x \right) \\
&\quad + \frac{1}{2}\int_{z+\mathsf{C}(2/|y|,R)} e^{-2\mathrm{i}\pi\, x\cdot y}\big(K(x) - K(x-z)\big)\,\mathrm{d}x.
\end{aligned}
$$

Since we have $z + \mathsf{C}(2/|y|, R) \subset \{|x| \;:\; |x| > 1/|y|\}$ and $1/|y| = 2|z|$ the last integral is bounded by $B/2$. Set

$$\mathsf{m}_{\varepsilon,R}^{(3)}(y) = \frac{1}{2}\left(\int_{\mathsf{C}(2/|y|,R)} e^{-2\mathrm{i}\pi\, x\cdot y} K(x)\,\mathrm{d}x - \int_{z+\mathsf{C}(2/|y|,R)} e^{-2\mathrm{i}\pi\, x\cdot y} K(x)\,\mathrm{d}x \right).$$

Since the symmetric difference of $\mathsf{C}(2/|y|, R)$ and $z + \mathsf{C}(2/|y|, R)$ is contained in $\mathsf{C}(1/|y|, 3/|y|) \cup \mathsf{C}(R - 1/|y|, R + 1/|y|)$, we have

$$
\begin{aligned}
2\left| \mathsf{m}_{\varepsilon,R}^{(3)}(y) \right| &\le \int_{\mathsf{C}(1/|y|,3/|y|)} |K(x)|\,\mathrm{d}x + \int_{\mathsf{C}(R-1/|y|,R+1/|y|)} |K(x)|\,\mathrm{d}x \\
&\le n\, \mathsf{v}_n B \left(\int_{1/|y|}^{3/|y|} r^{-1}\mathrm{d}r + \int_{R-1/|y|}^{R+1/|y|} r^{-1}\mathrm{d}r \right) \\
&\le 2n\, \mathsf{v}_n B \log 3.
\end{aligned}
$$

It results from those estimates that $\limsup_{R\nearrow\infty} |\mathsf{m}_{\varepsilon,R}(y)| \le CB$, where C depends only on n. Therefore $\|\widehat{K^{(\varepsilon)}}\|_\infty \le CB$.

Observe that, for $r > \varepsilon$, one has

$$\mathsf{m}_\varepsilon(y) = \int_{\varepsilon < |x| \le r} \big(e^{-2\mathrm{i}\pi\, x\cdot y} - 1\big) K(x)\,\mathrm{d}x + \mathsf{m}_r(y) \quad \text{for almost all } y.$$

It results that m_ε converges almost everywhere towards an L^∞-function which is denoted by m.

Take $\varphi \in \mathcal{S}$. One has

$$\left\langle \widehat{\operatorname{pv} K}, \varphi \right\rangle = \langle \operatorname{pv} K, \widehat{\varphi} \rangle$$

$$= \lim_{\varepsilon \searrow 0} \int K^{(\varepsilon)}(x)\, \widehat{\varphi}(x)\, \mathrm{d}x$$

$$= \lim_{\varepsilon \searrow 0} \int \widehat{K^{(\varepsilon)}}(x)\, \varphi(x)\, \mathrm{d}x$$

$$= \int \left(\lim_{\varepsilon \searrow 0} \widehat{K^{(\varepsilon)}}(x) \right) \varphi(x)\, \mathrm{d}x.$$

Let us now prove assertions 3 and 4. Consider

$$K_{\varepsilon,R}(x) = \begin{cases} K(x) & \text{if } \varepsilon < |x| < R, \\ 0 & \text{otherwise.} \end{cases}$$

This is an integrable function, so for $f \in L^2$ one has $\widehat{K_{\varepsilon,R} * f} = \mathrm{m}_{\varepsilon,R}\widehat{f}$. On the one hand, $K_{\varepsilon,R} * f$ converges uniformly towards $K^{(\varepsilon)} * f$ as R goes to ∞. On the other hand, when R goes to ∞, $\mathrm{m}_{\varepsilon,R}\widehat{f}$ converges in L^2 towards $\mathrm{m}_\varepsilon \widehat{f}$; therefore, due to the Plancherel theorem, $K_{\varepsilon,R} * f$ converges in L^2 to a limit which is the same as the uniform limit, namely, $K^{(\varepsilon)} * f$, since L^2-convergence implies the existence of a subsequence converging almost everywhere. So the Fourier transform of $K^{(\varepsilon)}$ is $\mathrm{m}_\varepsilon \widehat{f}$. A similar argument proves assertion 4.

2.4 ▪ Homogeneous kernels

As explained in Section B.2, homogeneous means positively homogeneous. So a function F on $\mathbb{R}^n$ is homogeneous of degree k (or k-homogeneous), if one has $F(tx) = t^k F(x)$ for all $t > 0$ and all x.

Lemma 2.6. *Let Ω be a function, integrable on the unit sphere S_{n-1}, considered as a homogeneous function of degree 0 (or 0-homogeneous) on $\mathbb{R}^n \setminus \{0\}$, (i.e., $\Omega(x) = \Omega(x/|x|)$ for all $x \neq 0$ in $\mathbb{R}^n$). We assume that*

$$\int_{S_{n-1}} \Omega(x)\, \mathrm{d}\sigma(x) = 0,$$

and, if Ω is not an odd function, we moreover assume that

$$\int_{S_{n-1}} |\Omega| \log^+ |\Omega| \mathrm{d}\sigma < \infty.$$

Set

$$\mathrm{m}_{\varepsilon,R}(y) = \int_{\varepsilon \leq |x| \leq R} \mathrm{e}^{-2\mathrm{i}\pi\, x \cdot y}\, \frac{\Omega(x)}{|x|^n}\, \mathrm{d}x.$$

Then

1. *$\mathrm{m}_{\varepsilon,R}$ is uniformly bounded;*

2. *for $y \neq 0$, the limit $\mathrm{m}(y) = \lim_{\varepsilon \searrow 0} \lim_{R \nearrow \infty} \mathrm{m}_{\varepsilon,R}(y)$ exists and*

$$\mathrm{m}(y) = \int_{S_{n-1}} \left(-\mathrm{i}\frac{\pi}{2} \operatorname{sgn}(x \cdot y) + \log \frac{|y|}{|x \cdot y|} \right) \Omega(x)\, \mathrm{d}\sigma(x).$$

Proof. Set $K(x) = \dfrac{\Omega(x)}{|x|^n}$. Due to hypothesis a, we have, for $y \neq 0$,

$$m_{\varepsilon,R}(y) = \int_{\varepsilon \leq |x| \leq R} \left(e^{-2i\pi\, x \cdot y} - \cos(2\pi\,|x||y|)\right) K(x)\, dx,$$

in which we write $m_{\varepsilon,R}(y) = v_{\varepsilon,R} - iu_{\varepsilon,R}$, where

$$
\begin{aligned}
u_{\varepsilon,R}(y) &= \int_{\varepsilon \leq |x| \leq R} \sin(2\pi\, x \cdot y)\, \Omega(x)\, \frac{dx}{|x|^n} \\
&= \int_{S_{n-1}} \Omega(x)\, d\sigma(x) \int_{\varepsilon}^{R} \frac{\sin(2\pi\, r\, x \cdot y)}{r}\, dr \\
&= \int_{S_{n-1}} \left(\int_{2\pi\varepsilon|x\cdot y|}^{2\pi R|x\cdot y|} \frac{\sin r}{r}\, dr \right) \operatorname{sgn}(x \cdot y)\, \Omega(x)\, d\sigma(x),
\end{aligned}
$$

and

$$
\begin{aligned}
v_{\varepsilon,R}(y) &= \int_{\varepsilon \leq |x| \leq R} \left(\cos(2\pi\, x \cdot y) - \cos(2\pi\,|x||y|)\right) K(x)\, dx \\
&= \int_{S_{n-1}} \Omega(x)\, d\sigma(x) \int_{\varepsilon}^{R} \left(\cos(2\pi r\,|x \cdot y|) - \cos(2\pi r|y|)\right) \frac{dr}{r} \\
&= \int_{S_{n-1}} \left(\int_{2\pi\varepsilon|x\cdot y|}^{2\pi\varepsilon|y|} \frac{\cos r}{r}\, dr - \int_{2\pi R|x\cdot y|}^{2\pi R|y|} \frac{\cos r}{r}\, dr \right) \Omega(x)\, d\sigma(x).
\end{aligned}
$$

Observe that this last term vanishes if Ω is odd.

Since $\displaystyle \sup_{0 < \alpha < \beta} \left| \int_{\alpha}^{\beta} \frac{\sin r}{r}\, dr \right| < \infty$ and $\displaystyle \int_{0}^{+\infty} \frac{\sin r}{r}\, dr = \frac{\pi}{2}$, we have

$$\lim_{\varepsilon \searrow 0} \lim_{R \to +\infty} u_{\varepsilon,R} = \frac{\pi}{2} \int_{S_{n-1}} \operatorname{sgn}(x \cdot y)\, \Omega(x)\, d\sigma(x).$$

To deal with the second term, observe first that

$$\left| \int_{\alpha}^{\beta} \frac{\cos r}{r}\, dr \right| \leq |\log \beta/\alpha|.$$

It results from the inequality $ab \leq (1 + 2a)\log(1 + 2a) + e^{b/2}$ (formula (2.6) in Exercise 2.7) that

$$
\begin{aligned}
\int_{S_{n-1}} \left| \Omega(x) \log \frac{|y|}{|x \cdot y|} \right| d\sigma(x) &\leq \int_{S_{n-1}} (1 + 2|\Omega(x)|)\log\big(1 + 2|\Omega(x)|\big)\, d\sigma(x) \\
&\quad + \int_{S_{n-1}} \left| \frac{|y|}{|x \cdot y|} \right|^{\frac{1}{2}} d\sigma(x) < +\infty. \qquad (2.2)
\end{aligned}
$$

Observe that, as $\displaystyle \int_{S_{n-1}} \left| \frac{|y|}{|x \cdot y|} \right|^{\frac{1}{2}} d\sigma(x)$ does not depend on y, the previous bound in (2.2) does not depend on y either.

So, for fixed $y \neq 0$,

$$\left| \left(\int_{2\pi\varepsilon|x\cdot y|}^{2\pi\varepsilon|y|} \frac{\cos r}{r}\,\mathrm{d}r - \int_{2\pi R|x\cdot y|}^{2\pi R|y|} \frac{\cos r}{r}\,\mathrm{d}r \right) \Omega(x) \right|$$

is dominated by a fixed integrable function (see (2.2)).

A first application of Lebesgue's dominated convergence yields

$$\lim_{R\to+\infty} v_{\varepsilon,R}(y) = \int_{S_{n-1}} \left(\int_{2\pi\varepsilon|x\cdot y|}^{2\pi\varepsilon|y|} \frac{\cos r}{r}\,\mathrm{d}r \right) \Omega(x)\,\mathrm{d}\sigma(x),$$

and a second one gives

$$\lim_{\varepsilon\to 0} \lim_{R\to+\infty} v_{\varepsilon,R}(y) = \int_{S_{n-1}} \Omega(x) \log\frac{|y|}{|x\cdot y|}\,\mathrm{d}\sigma(x).$$

So formula (2.3) is established and it is obvious that, for all $r > 0$ and $y \in \mathbb{R}^n \setminus \{0\}$, we have $\mathrm{m}(ry) = \mathrm{m}(y)$.

It also results from the above estimates that the functions $\mathrm{m}_{\varepsilon,R}$ and m are uniformly bounded.

Remark 2.7. *If we had replaced the condition* $\int_{S_{n-1}} |\Omega| \log^+ |\Omega|\,\mathrm{d}\sigma < +\infty$ *by the stronger one* $\int_{S_{n-1}} |\Omega|^p\mathrm{d}\sigma < +\infty$ *for some* $p \in (1,+\infty)$, *the bound (2.2) would be replaced by*

$$\int_{S_{n-1}} \left| \Omega(x) \log\frac{|y|}{|x\cdot y|} \right| \mathrm{d}\sigma(x) \leq \|\Omega\|_{L^p(\sigma)} \left(\int_{S_{n-1}} \left(\log\frac{|y|}{|x\cdot y|} \right)^{p'} \mathrm{d}\sigma(x) \right)^{1/p'},$$

which is obtained by using the Hölder inequality.

Theorem 2.8. *Let Ω be a function, integrable on the unit sphere S_{n-1}, considered as a homogeneous function of degree 0 on $\mathbb{R}^n \setminus \{0\}$ (i.e., $\Omega(x) = \Omega(x/|x|)$ for all $x \neq 0$ in $\mathbb{R}^n$). We assume that*

$$\int_{S_{n-1}} \Omega(x)\,\mathrm{d}\sigma(x) = 0,$$

and, if Ω is not an odd function, we moreover assume that

$$\int_{S_{n-1}} |\Omega| \log^+ |\Omega|\mathrm{d}\sigma < \infty.$$

Then the kernel $K(x) = \dfrac{\Omega(x)}{|x|^n}$ defines a tempered distribution whose Fourier transform is the function

$$\mathrm{m}(y) = \int_{S_{n-1}} \left(-\mathrm{i}\frac{\pi}{2}\,\mathrm{sgn}(x\cdot y) + \log\frac{|y|}{|x\cdot y|} \right) \Omega(x)\,\mathrm{d}\sigma(x). \qquad (2.3)$$

Moreover,

1. *the function* m *is a homogeneous of degree* 0 *and bounded;*

2. *for* $f \in L^2$, $K_{\varepsilon,R} * f$ *has a limit, when* R *goes to* $+\infty$, $T_\varepsilon f$ *in the space* L^2 *and* $\widehat{T_\varepsilon f} = \mathsf{m}_\varepsilon \widehat{f}$ *(with notation analogous to that of the preceding section);*

3. *for* $f \in L^2$, $T_\varepsilon f$ *has a limit* Tf *in* L^2 *when* ε *goes to* 0 *and* $\widehat{Tf} = \mathsf{m}\,\widehat{f}$.

Proof. The distribution pv K is well defined; see Exercise 2.2. Let $\varphi \in \mathscr{S}$. Let us compute $\langle K_{\varepsilon,R}, \widehat{\varphi} \rangle$. We have

$$\int_{\varepsilon<|x|<R} \frac{\Omega(x)}{|x|^n}\,\mathrm{d}x \int_{\mathbb{R}^n} e^{-2i\pi x \cdot y} \varphi(y)\,\mathrm{d}y = \int \varphi(y)\,\mathrm{d}y \int_{\varepsilon<|x|<R} e^{-2i\pi x \cdot y} \frac{\Omega(x)}{|x|^n}\,\mathrm{d}x$$

$$= \int \mathsf{m}_{\varepsilon,R}(y)\,\varphi(y)\,\mathrm{d}y.$$

As $K_{\varepsilon,R}$ converges in $\mathscr{S}'$ to pv K as R goes to ∞ and ε goes to 0, and since Lemma 2.6 allows dominated convergence, we get that the Fourier transform of pv K is m.

If $f \in L^2$, we have $\widehat{K_{\varepsilon,R} * f} = \mathsf{m}_{\varepsilon,R}\widehat{f}$. Again using dominated convergence, we get assertions 2 and 3.

In forthcoming chapters, it will be important to know whether kernels of the type considered in Theorem 2.8 fulfill the hypotheses of Theorem 2.5.

Proposition 2.9. *Let* Ω *be a function defined on the sphere* S_{n-1} *such that* $\int_0^1 \omega(\delta)\,\dfrac{\mathrm{d}\delta}{\delta}$
$< \infty$, *where* $\omega(\delta) = \sup\limits_{|x-y| \le \delta} |\Omega(x) - \Omega(y)|$. *We also assume that* $\int_{S_{n-1}} \Omega\,\mathrm{d}\sigma = 0$. *Then the kernel* $\dfrac{\Omega(x)}{|x|^n}$ *fulfills the hypotheses of Theorem* 2.5.

Proof. The first condition, called the Dini condition, implies $\lim_{\delta \to 0} \omega(\delta) = 0$, i.e., Ω is continuous, and thus bounded, on the sphere S_{n-1}. The kernel K obviously fulfills Hypothesis a of Theorem 2.5. We have

$$\int_{r_1<|x|\le r_2} K(x)\,\mathrm{d}x = \int_{r_1}^{r_2} r^{n-1}\mathrm{d}r \int_{S_{n-1}} \frac{\Omega(rx)}{r^n}\,\mathrm{d}\sigma(x)$$

$$= \int_{r_1}^{r_2} r^{-1}\mathrm{d}r \int_{S_{n-1}} \Omega(x)\,\mathrm{d}\sigma(x) = 0.$$

So Hypothesis c is fulfilled.

If $|x| \ge 2|y| > 0$, we have $\left| \dfrac{x-y}{|x-y|} - \dfrac{x}{|x|} \right| \le 2\dfrac{|y|}{|x|}$. Indeed, if θ stands for the angle between $x - y$ and x, we have $0 \le \theta < \pi/2$, $\sin\theta \le |y|/|x|$, and $\left| \dfrac{x-y}{|x-y|} - \dfrac{x}{|x|} \right| \le$

$2\sin\theta/2 \le 2\sin\theta$. So, we have $|\Omega(x-y) - \Omega(x)| \le \omega(2|y|/|x|)$. Under the same conditions, we have

$$\left| \frac{1}{|x-y|^n} - \frac{1}{|x|^n} \right| \le n2^{n+1} \frac{|y|}{|x|^{n+1}}$$

since the length of the gradient of $|.|^{-n}$ is bounded by $n2^{n+1}|x|^{-(n+1)}$ on the segment of line joining x to $x-y$.

We have, for $|x| \ge 2|y| > 0$,

$$|K(x-y) - K(x)| \le \frac{|\Omega(x-y) - \Omega(x)|}{|x|^n} + \left| \frac{1}{|x-y|^n} - \frac{1}{|x|^n} \right| |\Omega(x-y)|$$

$$\le \frac{1}{|x|^n} \omega\left(\frac{2|y|}{|x|} \right) + \frac{n2^{n+1}\|\Omega\|_\infty |y|}{|x|^{n+1}}. \tag{2.4}$$

Therefore

$$\int_{|x| \ge 2|y|} |K(x-y) - K(x)| \, \mathrm{d}x$$

$$\le n \, \mathsf{v}_n \int_{2|y|}^{+\infty} \omega(2|y|/r) \frac{\mathrm{d}r}{r} + n^2 2^{n+1} \mathsf{v}_n \|\Omega\|_\infty |y| \int_{2|y|}^{+\infty} \frac{\mathrm{d}r}{r^2}$$

$$= n \, \mathsf{v}_n \int_0^1 \frac{\omega(\delta) \, \mathrm{d}\delta}{\delta} + n^2 2^n \mathsf{v}_n \|\Omega\|_\infty.$$

So, the second hypothesis of Theorem 2.5 is fulfilled.

2.4.1 ▪ Hilbert and Riesz transforms

The Hilbert transform

The Hilbert transform H is the operator whose kernel is $1/x$ (on $\mathbb{R}$). The corresponding multiplier is $-\mathrm{i}\pi \operatorname{sgn}$.

The Hilbert operator is closely related to the partial sum operators. Let us set $\mathcal{H} = (\operatorname{Id} + \frac{\mathrm{i}}{\pi} H)/2$. Then, if $f \in L^2$,

$$\widehat{\mathcal{H}f}(y) = \begin{cases} \widehat{f}(y) & \text{if } y > 0, \\ 0 & \text{if } y < 0. \end{cases}$$

Let us consider the following operators acting on functions: $\varpi_t f(x) = \mathrm{e}^{2\mathrm{i}\pi t x} f(x)$ and $\check{C}f(x) = f(-x)$. These are isometries of L^2 (and also of all the L^p spaces, which will be useful later). It is easy to check that for $f \in L^2$ we have $\widehat{\varpi_t f}(y) = \widehat{f}(y-t)$ and $\widehat{\check{C}f}(y) = \widehat{f}(-y)$.

Consider the operator $S_R = (\check{C}\varpi_{-R}\mathcal{H}\varpi_R)^2$. It is bounded on L^2 with a bound independent of R. We can check that, for $f \in L^1 + L^2$,

$$\widehat{S_R f} = \widehat{f}\mathbf{1}_{[-R,R]}.$$

In other words,

$$S_R f(x) = \int_{-R}^{R} e^{2i\pi xy} \widehat{f}(y) \, \mathrm{d}y.$$

More generally, consider the operator $S_{[a,b]}$, where $a < b$, acting on L^2 so defined: $\widehat{S_{[a,b]} f} = \widehat{f} \mathbf{1}_{[a,b]}$. Then $S_{[a,b]} = \check{C} \varpi_{-b} \mathcal{H} \varpi_b \check{C} \varpi_{-a} \mathcal{H} \varpi_a$.

The Riesz transforms

Suppose $n \geq 2$. The kernel $\dfrac{c_n x_j}{(x_1^2 + x_2^2 + \cdots + x_n^2)^{\frac{n+1}{2}}}$, where c_n is a real constant which is defined below, is of the type considered in Theorem 2.8. It defines a bounded operator R_j on $L^2(\mathbb{R}^n)$ called the *jth Riesz transform*.

Let us study altogether the Riesz transforms by considering the vector kernel $\dfrac{c_n x}{|x|^{n+1}}$. This time the corresponding multiplier is a vector function, and, as previously stated, is so defined (for $y \neq 0$)

$$\mathsf{m}(y) = c_n \lim_{\substack{R \to \infty \\ \varepsilon \to 0}} \int_{\varepsilon \leq |x| \leq R} e^{-2i\pi \, x \cdot y} \frac{x}{|x|^{n+1}} \, \mathrm{d}x$$

$$= -i\, c_n \lim_{\substack{R \to \infty \\ \varepsilon \to 0}} \int_{\varepsilon \leq |x| \leq R} \sin(2\pi \, x \cdot y) \frac{x}{|x|^{n+1}} \, \mathrm{d}x.$$

Let $\rho \in SO(n)$ be a rotation. We have

$$\int_{\varepsilon \leq |x| \leq R} e^{-2i\pi \, x \cdot (\rho y)} \frac{x}{|x|^{n+1}} \, \mathrm{d}x = \int_{\varepsilon \leq |x| \leq R} e^{-2i\pi \, (\rho^{-1} x) \cdot y} \frac{x}{|x|^{n+1}} \, \mathrm{d}x$$

$$= \int_{\varepsilon \leq |x| \leq R} e^{-2i\pi \, x \cdot y} \frac{\rho x}{|x|^{n+1}} \, \mathrm{d}x.$$

As $i\,\mathsf{m}$ is obviously real, this last identity means that for any ρ and $y \neq 0$ we have $i\,\mathsf{m}(\rho y) = \rho(i\,\mathsf{m}(y))$.

We know that m is homogeneous of degree 0, so it is enough to compute it on the unit sphere.

So, $i\,\mathsf{m}(x)$ is invariant by any rotation which leaves x invariant. This implies that there exists $\lambda(x) \in \mathbb{R}$ such that $\mathsf{m}(x) = -i\lambda(x)x$. Now if x and y are two points in S_{n-1}, there exists a rotation ρ such that $y = \rho x$. We then have

$$\lambda(y)y = i\,\mathsf{m}(y) = i\,\mathsf{m}(\rho x) = \rho(i\,\mathsf{m}(x)) = \rho(\lambda(x)x) = \lambda(x)\rho x = \lambda(x)y.$$

Therefore λ is constant on S_{n-1}. The above constant c_n is taken so that $\mathsf{m}(y) = -i \dfrac{y}{|y|}$.

This means that for $f \in L^2$, one has $\widehat{R_j f}(y) = -i \dfrac{y_j}{|y|} \widehat{f}(y)$.

Let us compute the value of the constant c_n. According to Theorem 2.8, if m_n stands for the last component of m, then

$$\mathsf{m}_n(0, 0, \ldots, 0, 1) = -i\, c_n \frac{\pi}{2} \int_{S_{n-1}} |x_n| \, \mathrm{d}\sigma(x).$$

Therefore, $c_n = \frac{2}{\pi I}$, where $I = \int_{S_{n-1}} |x_n| \, \mathrm{d}\sigma(x)$.

One has

$$\int_{\mathbb{R}^n} |x_n| \, e^{-\pi |x|^2} \, dx = \int_{\mathbb{R}} |t| \, e^{-\pi t^2} \, dt = \frac{1}{\pi},$$

and also

$$\int_{\mathbb{R}^n} |x_n| \, e^{-\pi |x|^2} dx = \int_0^{+\infty} r^n e^{-\pi r^2} \left(\int_{S_{n-1}} |x_n| \, d\sigma(x) \right) dr$$

$$= \frac{\Gamma(\frac{n+1}{2})}{2 \, \pi^{\frac{n+1}{2}}} \, I.$$

It results that

$$c_n = \frac{\Gamma(\frac{n+1}{2})}{\pi^{\frac{n+1}{2}}}. \tag{2.5}$$

2.5 ▪ Exercises

Exercise 2.1. *Let T be the operator on L^2 defined by the multiplier* m. *Show that*

$$\|T\|_{2 \to 2} = \|\mathrm{m}\|_\infty.$$

Exercise 2.2. *Let Ω be an integrable function on the unit sphere S_{n-1} such that* $\int_{S_{n-1}} \Omega(x) \, d\sigma(x) = 0.$ *The function Ω is extended to $\mathbb{R}^n \setminus \{0\}$ as a 0-homogeneous function:* $\Omega(x) = \Omega(x/|x|)$ *for $x \neq 0$.*
   Prove that, for all $f \in \mathscr{S}(\mathbb{R}^n)$, the limit

$$\lim_{\varepsilon \to 0} \int_{|x| > \varepsilon} f(x) \, \frac{\Omega(x)}{|x|^n} \, dx$$

exists, and define a tempered distribution.

Exercise 2.3. *Let K be a C^1-function on $\mathbb{R}^n \setminus \{0\}$ such that*

$$\sup_{x \neq 0} |x|^{n+1} |\nabla K(x)| \leq 1.$$

Then there exists C such that

$$\sup_{y \neq 0} \int_{|x| \geq 2|y|} |K(x - y) - K(x)| \, dx \leq C < +\infty.$$

Exercise 2.4. *Let K be a locally integrable function on $\mathbb{R}^n \setminus \{0\}$ such that there exists B so that*

   1. for all $R > 0$, one has $\int_{|x| \leq R} |x| \, |K(x)| \, dx \leq BR;$

2. $\displaystyle\sup_{y\neq 0}\int_{|x|\geq 2|y|}|K(x-y)-K(x)|\,\mathrm{d}x \leq B$;

3. *for all r_1 and r_2 such that $0 < r_1 < r_2$, one has* $\displaystyle\left|\int_{r_1\leq|x|\leq r_2}K(x)\,\mathrm{d}x\right| \leq B$.

   Let $K_{\varepsilon,\eta}(x) = K(x)$ if $\varepsilon < |x| < \eta$ and 0 otherwise. Show that

   $$\sup_{0<\varepsilon<\eta<+\infty}\|\widehat{K_{\varepsilon,\eta}}\|_\infty \leq CB,$$

   where C depends only on the dimension n.

Exercise 2.5. *Let $y \in S_{n-1}$. Prove that, for all $\alpha < 1$,*

$$\int_{S_{n-1}}|x\cdot y|^{-\alpha}\mathrm{d}\sigma(x) < \infty.$$

Exercise 2.6. *Let $y \in S_{n-1}$. Prove that, for all $p > 0$,*

$$\int\left|\log\frac{1}{|x\cdot y|}\right|^p\,\mathrm{d}\sigma(x) < \infty.$$

Exercise 2.7 (Young's conjugate functions). *Let φ be a continuous increasing function from $[0, +\infty)$ onto $[0, +\infty)$. One sets*

$$\Phi(x) = \int_0^x \varphi(t)\,\mathrm{d}t \qquad and \qquad \Psi(y) = \int_0^y \varphi^{-1}(t)\,\mathrm{d}t.$$

1. *Prove that, for all $a \geq 0$ and $b \geq 0$, $ab \leq \Phi(a) + \Psi(b)$ and that the equality holds if and only if $b = \varphi(a)$.*

   Hint: *Draw a picture.*

2. *Prove the inequality, valid for nonnegative a and b,*

   $$ab \leq (a+1)\log(a+1) + e^b. \qquad (2.6)$$

   Hint: *Use the function $\varphi(t) = \log(1+t)$.*

Exercise 2.8. *Let Φ and Ψ be as in Exercise 2.7, and let f and g be two measurable functions.*

1. *Prove the inequality*

   $$\int|f(x)g(x)|\,\mathrm{d}x \leq \int \Phi(|f(x)|)\,\mathrm{d}x + \int \Psi(|g(x)|)\,\mathrm{d}x. \qquad (2.7)$$

2. *Define* $\|f\|_\Phi = \inf \left\{ t > 0 : \int \Phi(|f(x)|/t)\, dx \le 1 \right\}$. *Then*

$$\int |f(x)g(x)|\, dx \le \|f\|_\Phi \left(1 + \int \Psi(|g(x)|)\, dx \right)$$

and

$$\int |f(x)g(x)|\, dx \le 2\|f\|_\Phi \|g\|_\Psi.$$

3. *Prove that the set of f such that $\|f\|_\Phi < +\infty$ is a vector space. Such a space is called an* Orlicz *space.*

4. *Let f_j be a sequence of functions. Then $\displaystyle\lim_{j\to\infty} \int \Phi(t|f_j(x)|)\, dx = 0$, for all $t > 0$, if and only if $\displaystyle\lim_{j\to\infty} \|f_j\|_\Phi = 0$.*

Exercise 2.9. *Let $(X, \mathcal{A}, \mu)$ be a measure space. For a measurable function f the following conditions are equivalent:*

1. $f \in L^1(\mu)$ *and* $|f| \log^+ |f| \in L^1(\mu)$.

2. $|f| \log(2 + |f|) \in L^1(\mu)$.

Exercise 2.10. *Compute the Hilbert transform of $\dfrac{1}{\pi(1+x^2)}$.*

Exercise 2.11. *Let a and b be two numbers such that $a < b$. Compute the Hilbert transform* $\mathrm{pv} \displaystyle\int_{\mathbb{R}} \mathbf{1}_{[a,b]}(x - y)\, \dfrac{dy}{y}$ *of the indicator function of the interval $[a, b]$.*

Exercise 2.12. *Consider $2n$ real numbers $a_1 < b_1 < a_2 < b_2 < \cdots < a_n < b_n$ and set*

$$\varphi(x) = \log \prod_{1 \le j \le n} \frac{|x - a_j|}{|x - b_j|}.$$

a. *Prove that, for all $t \in \mathbb{R}$, the equation $\varphi(x) = t$ has at most $2n$ roots.*

b. *Show that, for all $t > 0$, the equation $\varphi(x) = t$ has $2n$ roots $\alpha_1, \dots, \alpha_n, \beta_1, \dots, \beta_n$ such that*

$$a_1 < \alpha_1 < b_1 < \beta_1 < a_2 < \alpha_2 < b_2 < \beta_2 < a_3 < \cdots < a_n < \alpha_n < b_n < \beta_n.$$

c. *Show that*

$$\sum \alpha_j = \frac{\sum a_j + e^t \sum b_j}{1 + e^t} \quad \text{and} \quad \sum \beta_j = \frac{\sum a_j - e^t \sum b_j}{1 - e^t}.$$

Exercise 2.13. *Let* $(a_j)_{1 \leq j \leq n}$ *and* $(b_j)_{1 \leq j \leq n}$ *be numbers as in Exercise 2.12. Consider the set* $F = \bigcup_{j=1}^{n} [a_j, b_j]$.

Show that, for all $t > 0$, *one has*

$$\lambda(\{|H\mathbf{1}_F| > t\}) = \frac{2\sum_{j=1}^{n}(b_j - a_j)}{\sinh t},$$

where λ *is the Lebesgue measure on* $\mathbb{R}$.

Exercise 2.14. *Let* E *be a Borel subset of* $\mathbb{R}$ *of finite Lebesgue measure. Show that, for all* $\varepsilon > 0$, *there exists a set* $F \subset \mathbb{R}$ *which is the union of finitely many bounded open intervals and such that* $\lambda(E \Delta F) \leq \varepsilon$, *where* $E \Delta F$ *stands for the symmetric difference of* E *and* F.

Exercise 2.15. *Show that, for any measurable subset* E *of* $\mathbb{R}$ *of finite Lebesgue measurable, one has*

$$\lambda(\{|H\mathbf{1}_E| > t\}) = \frac{2\lambda(E)}{\sinh t}$$

for all $t > 0$.

Chapter 3

The Calderón–Zygmund theory

3.1 ▪ The dyadic cubes

By a cube in $\mathbb{R}^n$ we mean a Cartesian product of intervals semiopen to the right and of the same length.

Let $Q = Q(a, \ell) = \prod_{j=1}^{n} [a_j, a_j + \ell)$ be a cube of side ℓ. The point $a = (a_j)$ is called the lower left corner of Q.

For $j \in \mathbb{Z}$, let $\mathcal{D}_j(Q)$ stand for the collection of cubes of side $2^{-j}\ell$ whose lower left corners are the points of the lattice $a + 2^{-j}\ell\mathbb{Z}$. These cubes are the *dyadic cubes of generation j subordinate to Q*. Let $\mathcal{D}(Q) = \bigcup_{j \in \mathbb{Z}} \mathcal{D}_j(Q)$. In the standard case when Q is the unit cube $Q(0, 1)$, one just says dyadic cubes.

When the context makes it clear, we write $\mathcal{D}$ and $\mathcal{D}_j$ instead of $\mathcal{D}(Q)$ and $\mathcal{D}_j(Q)$.

A maximal operator is associated with any family $\mathcal{D}$ of dyadic cubes:

$$\mathsf{M}_{\mathcal{D}} f(x) = \sup_{x \in Q \subset \mathcal{D}} \frac{1}{|Q|} \int_Q |f(y)| \, \mathrm{d}y. \tag{3.1}$$

One obviously has $\mathsf{M}_{\mathcal{D}} f \leq n^{\frac{n}{2}} \mathsf{v}_n \mathsf{M} f$ (a cube of side ℓ containing x is contained in the ball $\mathsf{B}(x, \ell\sqrt{n})$). So the operator $\mathsf{M}_{\mathcal{D}}$ is weak L^1 and bounded on any L^p for $1 < p \leq \infty$. Also for $f \in L^1_{\mathrm{loc}}$ and for almost every x (in particular at any Lebesgue point of f), one has

$$\lim_{j \to \infty} \frac{1}{|Q_j(x)|} \int_{Q_j(x)} |f(y) - f(x)| \, \mathrm{d}y = 0,$$

where $Q_j(x)$ is the element of $\mathcal{D}_j$ which contains x.

This comparison to the Hardy–Littlewood maximal operator takes advantage of the work previously done. Nevertheless a direct proof is not only possible, but also better.

We begin with an observation which is trite but powerful.

Remark 3.1. *If Q_1 and Q_2 are elements of $\mathcal{D} = \bigcup_{j \in \mathbb{Z}} \mathcal{D}_j$, either they are disjoint or one contains the other.*

27

Let $f \in L^1$ and $t > 0$. Consider the set $\mathscr{E}$ of cubes $Q \in \mathcal{D}$ such that $\dfrac{1}{|Q|} \displaystyle\int_Q |f(x)| \, \mathrm{d}x > t$.

Let $Q_1, Q_2, \ldots, Q_m, \ldots$ be the elements of $\mathscr{E}$ maximal with respect to inclusion. They are disjoint and any element of $\mathscr{E}$ is contained in one of them. We have $\{M_{\mathcal{D}} f > t\} = \bigcup_j Q_j$, so

$$|\{M_{\mathcal{D}} f > t\}| = \sum_j |Q_j| \le \frac{1}{t} \int_{Q_j} |f(x)| \, \mathrm{d}x$$

$$= \frac{1}{t} \int_{\bigcup Q_j} |f(x)| \, \mathrm{d}x \le \frac{\|f\|_1}{t}.$$

Remark 3.2. *This proof can be explained in terms of martingale and stopping time.*

3.2 ▪ The Calderón–Zygmund decomposition

Lemma 3.3. *Let f be an element of $L^1(\mathbb{R}^n)$ and let t be a positive number. Then there exist countably many cubes $Q_1, Q_2, \ldots$ which are disjoint and such that*

1. *$|f(x)| \le t$ for almost all x in the set $F = {}^c\bigcup_{j \ge 1} Q_j$;*

2. *for any j, one has $t < \dfrac{1}{|Q_j|} \displaystyle\int_{Q_j} |f(x)| \, \mathrm{d}x \le 2^n t$.*

Proof. First choose a system of dyadic cubes $\mathcal{D}$, then fix $t > 0$.

Let $\mathscr{E}$ be the set of cubes $Q \in \mathcal{D}$ such that

$$\frac{1}{|Q|} \int_Q |f(x)| \, \mathrm{d}x > t. \tag{3.2}$$

Let $Q_1, Q_2, \ldots, Q_n, \ldots$ be the elements of $\mathscr{E}$ maximal with respect to inclusion. Then, as previously stated, these cubes are disjoint and any element of $\mathscr{E}$ is contained in one of the Q_j. Set $F = {}^c\bigcup_{j \ge 1} Q_j$. If $x \in F$, then, for all n, x belongs to a cube $Q'_n \in \mathcal{D}_n$ such that $\dfrac{1}{|Q'_n|} \displaystyle\int_{Q'_n} |f(y)| \, \mathrm{d}y \le t$. Then it results from the Lebesgue differentiation theorem that $|f(x)| \le t$ for almost every x in F.

Let $\widetilde{Q}_j$ be the mother cube of Q_j (i.e., the cube in $\mathcal{D}$, of side twice that of Q_j, which contains Q_j). Due to the maximality of Q_j, one has $\displaystyle\int_{\widetilde{Q}_j} |f(x)| \, \mathrm{d}x \le t \, |\widetilde{Q}_j|$ and

$$\frac{1}{|Q_j|} \int_{Q_j} |f(x)| \, \mathrm{d}x = \frac{2^n}{|\widetilde{Q}_j|} \int_{Q_j} |f(x)| \, \mathrm{d}x \le \frac{2^n}{|\widetilde{Q}_j|} \int_{\widetilde{Q}_j} |f(x)| \, \mathrm{d}x \le 2^n t.$$

Lemma 3.4 (Calderón–Zygmund decomposition). *Let f be an integrable function on $\mathbb{R}^n$ and let t be a positive number. Then there exist countably many disjoint cubes $Q_1, Q_2, \ldots$ so that*

1. *$|f(x)| \le t$ for almost all x in the set $F = {}^c\bigcup_{j \ge 1} Q_j$;*

2. *for any j, one has* $\dfrac{1}{|Q_j|} \displaystyle\int_{Q_j} |f(x)|\,dx \le 2^n t;$

3. $|\Omega| \le \dfrac{1}{t}\,\|f\|_1,$ *where* $\Omega = \bigcup Q_j.$

Proof. We use the notation of Lemma 3.3. By the second conclusion of this lemma, and since the cubes Q_j are disjoint, we have

$$|\Omega| = \sum_j |Q_j| \le \sum_j \frac{1}{t}\int_{Q_j}|f(x)|\,dx \le \frac{1}{t}\int_{\bigcup Q_j}|f(x)|\,dx \le \frac{1}{t}\,\|f\|_1.$$

3.3 ▪ Singular integrals

Theorem 3.5 (liminal theorem). *Let $K \in L^2(\mathbb{R}^n)$. Assume there exists B such that*

a. $\|\widehat{K}\|_\infty \le B;$

b. $\displaystyle\sup_{y \ne 0}\int_{|x| \ge 2|y|} |K(x-y) - K(x)|\,dx \le B;$

If $f \in L^1(\mathbb{R}^n)+L^2(\mathbb{R}^n)$, one sets $Tf = K * f$. Then we have the following conclusions:

1. *There exists $C > 0$ depending only on n and B such that, for all $f \in L^1(\mathbb{R}^n)$ and for all $t > 0$, one has*

$$|\{|Tf| > t\}| \le \frac{C}{t}\,\|f\|_1.$$

2. *For all $p \in (1, +\infty)$ there exists $C_p > 0$ depending only on p, n, and B such that, for any $f \in L^2(\mathbb{R}^n) \cap L^p(\mathbb{R}^n)$, one has $\|Tf\|_p \le C_p\|f\|_p$.*

Proof. We begin with a few observations. Due to properties of convolution, if f is in L^1, $Tf(x)$ is defined almost everywhere and is in L^2, and, if $f \in L^2$, $Tf(x)$ is everywhere defined and is a continuous function vanishing at infinity. Therefore, Tf is well defined for $f \in L^1 + L^2$. If $f \in L^2$, one has $\widehat{Tf} = \widehat{K}\,\widehat{f}$, so

$$\|Tf\|_2 = \|\widehat{Tf}\|_2 \le B\,\|\widehat{f}\|_2 = B\,\|f\|_2. \tag{3.3}$$

Proof of assertion 1.

We use the notation of Lemma 3.4 and set

$$g(x) = \begin{cases} f(x) & \text{if } x \in F, \\ \dfrac{1}{|Q_j|}\displaystyle\int_{Q_j} f(y)\,dy & \text{if } x \in Q_j, \end{cases}$$

$$b(x) = \begin{cases} 0 & \text{if } x \in F, \\ f(x) - \dfrac{1}{|Q_j|}\displaystyle\int_{Q_j} f(y)\,dy & \text{if } x \in Q_j, \end{cases}$$

so that we have $f = g + b$.

As usual, we write $|\{|Tf| > t\}| \leq |\{|Tg| > t/2\}| + |\{|Tb| > t/2\}|$.
We have

$$\|g\|_2^2 = \int_F |f(x)|^2 \mathrm{d}x + \sum_j |\mathrm{Q}_j| \left(\frac{1}{|\mathrm{Q}_j|} \int_{\mathrm{Q}_j} f(x)\, \mathrm{d}x \right)^2$$

$$\leq t \int_F |f(x)|\, \mathrm{d}x + 2^{2n} t^2 |\Omega|$$

$$\leq (1 + 2^{2n})\, t\, \|f\|_1,$$

which implies

$$|\{|Tg| > t/2\}| \leq \frac{4}{t^2} \|Tg\|_2^2 \leq \frac{4(1 + 2^{2n})B^2}{t} \|f\|_1.$$

Now, we proceed to estimate $|\{|Tb| > t/2\}|$. If Q is a cube, let Q^* stand for the cube which has the same center as Q and whose side is $2\sqrt{n}$ times that of Q. Set $\Omega^* = \bigcup \mathrm{Q}_j^*$ and $F^* = {}^c\Omega^*$. We have

$$|\Omega^*| \leq \sum |\mathrm{Q}_j^*| = 2^n n^{n/2} \sum |\mathrm{Q}_j| \leq \frac{2^n n^{n/2}}{t} \|f\|_1.$$

Therefore,

$$|\{|Tb| > t/2\}| \leq \frac{2^n n^{n/2}}{t} \|f\|_1 + |F^* \cap \{|Tb| > t/2\}|$$

$$\leq \frac{2^n n^{n/2}}{t} \|f\|_1 + \frac{2}{t} \int_{F^*} |Tb(x)|\, \mathrm{d}x.$$

To end the proof, we have to show that $\displaystyle\int_{F^*} |Tb(x)|\, \mathrm{d}x$ is bounded. Write $b = \sum b_j$, where $b_j = \left(f - \dfrac{1}{|\mathrm{Q}_j|} \displaystyle\int_{\mathrm{Q}_j} f(x)\, \mathrm{d}x \right) \mathbf{1}_{\mathrm{Q}_j}$. Obviously, $\displaystyle\int |b_j(x)|\, \mathrm{d}x \leq 2^{n+1} t\, |\mathrm{Q}_j|$. Let $y^{(j)}$ be the center of Q_j. Note that

$$Tb_j(x) = \int_{\mathrm{Q}_j} \left(K(x - y) - K(x - y^{(j)}) \right) b_j(y)\, \mathrm{d}y$$

since $\displaystyle\int_{\mathrm{Q}_j} b_j(x)\, \mathrm{d}x = 0$.

We have

$$\int_{F^*} |Tb_j(x)|\, \mathrm{d}x \leq \int_{F^*} \mathrm{d}x \int_{\mathrm{Q}_j} \left| K(x - y) - K(x - y^{(j)}) \right| |b_j(y)|\, \mathrm{d}y$$

$$\leq \int_{{}^c\mathrm{Q}_j^*} \mathrm{d}x \int_{\mathrm{Q}_j} \left| K(x - y) - K(x - y^{(j)}) \right| |b_j(y)|\, \mathrm{d}y$$

$$= \int_{\mathrm{Q}_j} |b_j(y)|\, \mathrm{d}y \int_{{}^c\mathrm{Q}_j^*} \left| K(x - y) - K(x - y^{(j)}) \right| \mathrm{d}x$$

$$\leq 2^{n+1} t\, |\mathrm{Q}_j| \sup_{y \in \mathrm{Q}_j} \int_{{}^c\mathrm{Q}_j^*} \left| K(x - y) - K(x - y^{(j)}) \right| \mathrm{d}x.$$

But, due to hypothesis 2 of Theorem 3.5, we have

$$\int_{{}^cQ_j^*} \left| K(x-y) - K(x - y^{(j)}) \right| \, \mathrm{d}x$$

$$= \int_{{}^cQ_j^*} \left| K\left(x - y^{(j)} - (y - y^{(j)}) \right) - K(x - y^{(j)}) \right| \, \mathrm{d}x$$

$$\leq \int_{|x - y^{(j)}| \geq 2|y - y^{(j)}|} \left| K\left(x - y^{(j)} - (y - y^j) \right) - K(x - y^{(j)}) \right| \, \mathrm{d}x \leq B.$$

Finally, we get

$$\int_{F^*} |Tb(x)| \, \mathrm{d}x \leq \sum \int_{F^*} |Tb_j(x)| \, \mathrm{d}x \leq 2^{n+1} t \, B \sum |Q_j| \leq 2^{n+1} B \, \|f\|_1. \quad (3.4)$$

Proof of assertion 2.

Since the operator T is bounded on L^2 and of weak type $(1,1)$, the Marcinkiewicz interpolation theorem (Theorem 8.1) gives the result when $1 < p < 2$.

The operator T commutes with translations and is bounded on L^p for $1 < p \leq 2$, and Lemma 2.4 shows that it is also bounded on L^p for $2 \leq p < +\infty$.

Comments

- When $p > 2$, the convolution $K * f$ is not a priori pointwise defined for all $f \in L^p$. Nevertheless, it is defined on the vector space $L^2 \cap L^p$, which is dense in L^p, and is continuous with respect to the norm $\| \; \|_p$, so it can be extended by continuity to the whole space L^p and defines a bounded operator on L^p.

- The exact values of the constants occurring in the above inequalities are meaningless for our purpose. In what follows, C stands for such constants, and different occurrences of C could refer to different values.

- When K is C^1 except maybe at the origin, the second hypothesis of Theorem 3.5 is fulfilled if the following condition holds (see the next lemma):

  There exists C such that, for all $x \in \mathbb{R}^n \setminus \{0\}$, one has $|\nabla K(x)| \leq C \, |x|^{-n-1}$,

  where ∇K stands for the gradient of K.

Lemma 3.6. *Let K be a C^1-function on $\mathbb{R}^n \setminus \{0\}$. If*

$$\sup |x|^{n+1} |\nabla K(x)| = B < +\infty,$$

then

$$\sup_{y \neq 0} \int_{|x| \geq 2|y|} |K(x - y) - K(x)| \, \mathrm{d}x \leq 2^n n v_n B.$$

Proof. For $0 \leq t \leq 1$, we have $|x - ty| \geq |x| - |y| \geq |x/2|$; therefore $|\nabla K(z)| \leq 2^{n+1} |x|^{-(n+1)} B$ for all z on the line between x and $x - y$. Then

$$\int_{|x| \geq 2|y|} |K(x - y) - K(x)| \, \mathrm{d}x \leq 2^{n+1} B \int_{|x| \geq 2|y|} \frac{|y|}{|x|^{n+1}} \, \mathrm{d}x = 2^n n v_n B.$$

Since the bounds for the operator norm on L^p does not depend on the L^2 norm of K, it is likely that we can get rid of the hypothesis that K is square integrable. This is the purpose of the next theorem.

Theorem 3.7. *Let K be a locally integrable function on $\mathbb{R}^n \setminus \{0\}$ such that there exists B so that*

a. $|K(x)| \leq B |x|^{-n}$;

b. $\displaystyle\sup_{y \neq 0} \int_{|x| \geq 2|y|} |K(x - y) - K(x)| \, dx \leq B$;

c. $\displaystyle\int_{r_1 \leq |x| \leq r_2} K(x) \, dx = 0.$

For $f \in \bigcup_{p \geq 1} L^p(\mathbb{R}^n)$ and $\varepsilon > 0$, one sets

$$T_\varepsilon f(x) = \int_{|y| > \varepsilon} K(y) \, f(x - y) \, dy.$$

Then

1. *for all $1 < p < \infty$, there exists C_p depending only on n, p, and B such that, for all f and all ε, one has $\|T_\varepsilon f\|_p \leq C_p \|f\|_p$;*

2. *if $f \in L^p$ (with $1 < p < +\infty$), $T_\varepsilon f$ has a limit Tf in L^p as ε goes to 0, and $\|Tf\|_p \leq C_p \|f\|_p$.*

Proof. First, observe that $T_\varepsilon f$ is well defined: for $p' > 1$, $|x|^{-np'}$ is integrable outside the ball $B(0, \varepsilon)$, so, for $f \in L^p$ ($1 \leq p < +\infty$), $T_\varepsilon f$ is a continuous function.

Proof of assertion 1.

Set $K^{(\varepsilon)}(x) = K(x)$ if $|x| > \varepsilon$, $K^{(\varepsilon)}(x) = 0$ if $|x| \leq \varepsilon$.

To prove this assertion it is enough to prove that the kernel $K^{(\varepsilon)}$ fulfills the hypotheses of Theorem 3.5 with constants depending on n and B, but not on ε.

Theorem 2.5 tells us that $\widehat{K^{(\varepsilon)}}$ is uniformly bounded. So, we only have to show that

$$\sup_{y \neq 0} \int_{|x| \geq 2|y|} |K^{(\varepsilon)}(x - y) - K^{(\varepsilon)}(x)| \, dx$$

is bounded independently on ε.

First, consider, for $y \neq 0$, the quantity

$$A = \int_{|x| > 2|y|} \left| K^{(\varepsilon)}(x - y) - K^{(\varepsilon)}(x) \right| \, dx.$$

One has

$$A = \int_{\substack{|x| > 2|y| \\ |x - y| > \varepsilon \\ |x| > \varepsilon}} |K(x - y) - K(x)| \, dx$$

$$+ \int_{\substack{|x| > 2|y| \\ |x - y| > \varepsilon \\ |x| \leq \varepsilon}} |K(x - y)| \, dx + \int_{\substack{|x| > 2|y| \\ |x - y| \leq \varepsilon \\ |x| > \varepsilon}} |K(x)| \, dx,$$

so

$$A \leq B + \int_{\varepsilon < |x-y| \leq 3\varepsilon/2} |K(x-y)|\, \mathrm{d}x + \int_{\varepsilon < |x| \leq 2\varepsilon} |K(x)|\, \mathrm{d}x$$

$$\leq B + 2 \int_{\varepsilon < |x| \leq 2\varepsilon} \frac{B}{|x|^n}\, \mathrm{d}x = (1 + 2n\, \mathsf{v}_n \log 2)\, B,$$

where v_n stands for the volume of the unit ball.

The uniform estimate of $\widehat{K^{(\varepsilon)}}$ results from Theorem 2.5.

Proof of assertion 2.

If φ is C^1 with compact support, due to Hypothesis c of Theorem 3.7, we have

$$T_\varepsilon \varphi(x) = \int_{\varepsilon < |y| < 1} K(y)\, (\varphi(x-y) - \varphi(x))\, \mathrm{d}y + \int_{|y| > 1} K(y)\, \varphi(x-y)\, \mathrm{d}y.$$

But $|\varphi(x-y) - \varphi(x)| \leq |y|\, \|\nabla\varphi\|_\infty$, so

$$\left| K(y)\, (\varphi(x-y) - \varphi(x)) \right| \leq B\, \|\nabla\varphi\|_\infty |y|^{1-n}.$$

It results that the function $\displaystyle\int_{\varepsilon < |y| < 1} K(y)\, (\varphi(x-y) - \varphi(x))\, \mathrm{d}y$, whose support is contained in a fixed compact set, converges uniformly, and therefore in L^p, towards $\displaystyle\int_{|y| < 1} K(y)\, (\varphi(x-y) - \varphi(x))\, \mathrm{d}y$ as ε goes to 0. We have just proved that $T_\varepsilon \varphi$ converges in L^p as ε goes to 0.

Now take $f \in L^p$. For any φ as above, we have

$$\|T_\varepsilon f - T_{\varepsilon'} f\|_p \leq \|T_\varepsilon \varphi - T_{\varepsilon'}\varphi\|_p + \|T_\varepsilon(f - \varphi)\|_p + \|T_{\varepsilon'}(f - \varphi)\|_p,$$

which implies

$$\lim_{\eta \searrow 0} \sup_{\substack{\varepsilon < \eta \\ \varepsilon' < \eta}} \|T_\varepsilon f - T_{\varepsilon'} f\|_p \leq 2C_p \|f - \varphi\|_p.$$

But, as C^1-functions with compact support are dense in L^p, the right-hand side of the last inequality can be arbitrarily small. This means that $T_\varepsilon f$ is Cauchy.

Here is an important consequence of this theorem.

Theorem 3.8. *Let Ω be a complex function on $\mathbb{R}^n$, homogeneous of degree 0 (i.e, $\Omega(tx) = \Omega(x)$ for all $t > 0$ and $x \in \mathbb{R}^n$) such that*

1. $\displaystyle\int_{S_{n-1}} \Omega(x)\, \mathrm{d}\sigma(x) = 0;$

2. $\displaystyle\int_0^1 \omega(\delta)\, \frac{\mathrm{d}\delta}{\delta} < \infty$, *where* $\omega(\delta) = \sup_{\substack{|x|=|y|=1 \\ |x-y|<\delta}} |\Omega(x) - \Omega(y)|.$

Then the kernel $K(x) = \dfrac{\Omega(x)}{|x|^n}$ fulfills the hypotheses of Theorem 3.7 and, if T stands for the corresponding operator, one has

$$\widehat{Tf} = \mathsf{m}\, \widehat{f}, \quad \text{for } f \in L^2,$$

where m *is homogeneous of degree 0 and*

$$\mathsf{m}(y) = \int_{S_{n-1}} \left(-\mathrm{i}\,\frac{\pi}{2}\,\mathrm{sgn}(x \cdot y) + \log\frac{|y|}{|x \cdot y|} \right) \Omega(x)\,\mathrm{d}\sigma(x).$$

Proof. See Proposition 2.9 and Theorem 2.8.

3.3.1 ▪ Almost everywhere convergence

First, we state two lemmas.

Lemma 3.9. *Let T be an operator of the type considered in Theorem 3.7. If $1 < p < \infty$, $f \in L^p$, and $g \in \mathscr{S}$, then*

$$Tf * g = Tg * f.$$

Proof. We know that Tf and Tg are the limits in L^p and $L^{p'}$ of $T_\varepsilon f = K^{(\varepsilon)} * f$ and $T_\varepsilon g = K^{(\varepsilon)} * g$ (these convolutions are well defined because $K^{(\varepsilon)}$ is in $L^{p'}$ as well as in L^p) as ε goes to 0. But we have, due to elementary properties of convolution, $(K^{(\varepsilon)} * f) * g = (f * K^{(\varepsilon)}) * g = f * (K^{(\varepsilon)} * g)$. So $T_\varepsilon f * g = T_\varepsilon g * f$, and we can take the limit as ε goes to 0.

Lemma 3.10. *Let K be as in Theorem 3.8 and, as usual, $f_\eta(x) = \eta^{-n} f(x/\eta)$. Then*

1. $(K^{(1)})_\eta = K^{(\eta)}$;

2. *if $f \in L^p$ for some $p \in (1, +\infty)$, one has $(Tf)_\eta = T(f_\eta)$.*

Proof. Assertion 1 comes from the homogeneity of K. To prove assertion 2, write

$$
\begin{aligned}
T_\varepsilon(f_\eta)(x) &= \int_{|y|>\varepsilon|} \eta^{-n} f\big(\eta^{-1}(x - y)\big)\,K(y)\,\mathrm{d}y \\
&= \int_{|z|>\varepsilon/\eta} f(\eta^{-1}x - z)\,K(\eta z)\,\mathrm{d}z \\
&= \int_{|z|>\varepsilon/\eta} f(\eta^{-1}x - z)\,\eta^{-n}K(z)\,\mathrm{d}z = \eta^{-n}\big(T_{\varepsilon/\eta}f\big)(\eta^{-1}x).
\end{aligned}
$$

The conclusion follows by taking the limits as ε goes to 0.

Lemma 3.11. *With the same hypotheses and notation as in Theorem 3.8, set*

$$T_\varepsilon f(x) = \int_{|y|>\varepsilon} K(y)\,f(x - y)\,\mathrm{d}y \qquad \text{and} \qquad T^* f(x) = \sup_{\varepsilon>0} |T_\varepsilon f(x)|.$$

Then there exists C such that, for all $f \in \bigcup_{p>1} L^p$,

$$T^* f \le \mathsf{M}(Tf) + C\mathsf{M}f.$$

Proof. Let φ be a nonnegative radial C^1-function supported in the unit ball $\mathsf{B}\,(0, 1)$, of integral 1, and such that $\varphi(x)$ is a nonincreasing function of $|x|$.

According to Section 2.2 the convolution $T\varphi$ is defined as a principal value: when $|x| \le 2$, we have

$$T\varphi(x) = \lim_{\varepsilon \searrow 0} \int_{|y| > \varepsilon} K(y)\,\varphi(x-y)\,\mathrm{d}y = \int_{|y| \le 3} K(y)\,(\varphi(x-y) - \varphi(x))\,\mathrm{d}y.$$

It results that $\sup_{|x| \le 2} \left| T\varphi(x) - K^{(1)}(x) \right| < \infty.$

Consider the function $\Phi = T\varphi - K^{(1)}$. When $|x| > 2$,

$$\Phi(x) = \int_{|y| \le 1} \left(K(x-y) - K(x) \right) \varphi(y)\,\mathrm{d}y.$$

By using (2.4), we get

$$|\Phi(x)| \le \frac{1}{|x|^n}\,\omega\left(\frac{2}{|x|}\right) + \frac{n2^{n+1}\|\Omega\|_\infty}{|x|^{n+1}}.$$

So, $x \mapsto \sup_{|y| \ge |x|} |\Phi(y)|$ is integrable, and we can use Proposition 1.9. There exists C, independent of ε, such that $|\Phi_\varepsilon * f| \le C\mathsf{M}f$ for all $f \in L^1_{\mathrm{loc}}$.

Due to Lemma 3.10, $\Phi_\varepsilon = T(\varphi_\varepsilon) - K^{(\varepsilon)}$. So $T_\varepsilon f = (T\varphi_\varepsilon) * f - \Phi_\varepsilon * f$. We know from Lemma 3.9 that, for $f \in \bigcup_{p>1} L^p$, we have $(T\varphi_\varepsilon) * f = Tf * \varphi_\varepsilon$. Then we obtain $|T_\varepsilon f| \le \mathsf{M}(Tf) + C\mathsf{M}f$.

Theorem 3.12. *With the same notation as in Lemma* 3.11 *one has*

1. *for $p \in (1, +\infty)$, T^* is bounded on L^p;*

2. *T^* is of weak type $(1,1)$;*

3. *$\lim_{\varepsilon \searrow 0} T_\varepsilon f(x)$ exists almost everywhere ($f \in L^p$, $1 < p < +\infty$).*

The first assertion is a consequence of Lemma 3.11 and Theorems 3.7 and 1.2.

To prove the second assertion we revisit the proof of Theorem 3.5. First, we need a simple geometric lemma.

Lemma 3.13. *Let Q be a cube, Q^* be the "doubled cube," $x \notin \mathsf{Q}^*$, $z \in \mathsf{Q}$, and $\rho = |x - z|$. Then, there exist α an β such that for all $y \in \mathsf{Q}$, $\alpha\rho \le |x - y| \le \beta\rho$.*

Proof. The diameter of Q is $\ell\sqrt{n}$, and that of Q^* is $2\ell\sqrt{n}$. We always have $\rho \ge \ell/2$.

We have $|x - y| \le \rho + \ell\sqrt{n} \le \rho(1 + 2\sqrt{n})$. So, we can take $\beta = 1 + 2\sqrt{n}$.

Let γ be positive number to be chosen later. If $\rho \le \ell/\gamma$, then $|x-y| \ge \ell/2 \ge \gamma\rho/2$.

If $\rho \ge \ell/\gamma$, then $|x - y| \ge \rho - \ell\sqrt{n} \ge (1 - \gamma\sqrt{n})\rho$.

If we choose γ such that $\gamma/2 = 1 - \gamma\sqrt{n}$, i.e., $\gamma = 2/(1 + 2\sqrt{n})$, we see that α can be taken as $1/(1 + 2\sqrt{n})$.

We have to estimate $|\{T^*f > t\}|$. As previously, we write $f = g + b$, with $b = \sum b_j$. The term $|\{T^*g > t/2\}|$ causes no problem, as we know that T^* is bounded on L^2.

Fix $x \in F^*$ and ε. Let us compute

$$T_\varepsilon b_j(x) = \int_{Q_j} K^{(\varepsilon)}(x-y)\, b_j(y)\, \mathrm{d}y.$$

If $Q_j \subset B(x,\varepsilon)$, then $T_\varepsilon b_j(x) = 0$. If $Q_j \subset \{y : |x-y| > \varepsilon\}$, then $T_\varepsilon b_j(x) = \int_{Q_j}(K(x-y) - K(x-y^{(j)}))\, b_j(y)\, \mathrm{d}y$. Otherwise $Q_j \cap B(x,\varepsilon) \neq \emptyset$; then due to Lemma 3.11, $Q_j \subset B(x, \beta\varepsilon)$ and we have

$$|T_\varepsilon b_j(x)| \leq (\beta\varepsilon)^{-n} \|\Omega\|_\infty \int_{Q_j} |b_j(y)|\, \mathrm{d}y \leq 2\,(\alpha\varepsilon)^{-n} \|\Omega\|_\infty \int_{Q_j} |f(y)|\, \mathrm{d}y.$$

By summing these estimates we get

$$|T_\varepsilon b(x)| \leq \sum_j \int_{Q_j} |(K(x-y) - K(x-y^{(j)}))\, b_j(y)|\, \mathrm{d}y$$
$$+ 2\,(\alpha\varepsilon)^{-n} \|\Omega\|_\infty \int_{|x-y| \leq 4\varepsilon} |f(y)|\, \mathrm{d}y$$
$$\leq \sum_j \int_{Q_j} |(K(x-y) - K(x-y^{(j)}))\, b_j(y)|\, \mathrm{d}y + C\,\mathsf{M}f(x).$$

Indeed, as the right-hand side does not depend on ε, we get an upper bound for $T^*b(x)$:

$$T^*b(x) \leq S(x) + C\,\mathsf{M}f(x),$$

where

$$S(x) = \sum_j \int_{Q_j} |(K(x-y) - K(x-y^{(j)}))\, b_j(y)|\, \mathrm{d}y.$$

By arguing as for proving (3.4) (to be more precise, by repeating the same calculations done to evaluate $\int_{F^*} |Tb_j(s)|\, \mathrm{d}x$) we get

$$\int_{{}^c\Omega^*} S(x)\, \mathrm{d}x \leq 2^n t\, B \sum |Q_j| \leq 2^n B\, \|f\|_1.$$

Then we write

$$\{T^*f > t\} \subset \{T^*g > t/2\} \cup \{\mathsf{M}f > t/4\} \cup \{x \notin \Omega^* : S(x) > t/4\} \cup \Omega^*$$

to end the proof.

Once we have this maximal inequality, the almost everywhere convergence follows as in the proof of the Lebesgue differentiation theorem (Theorem 1.4).

3.4 ▪ Exercises

Exercise 3.1. *Let f be an integrable function on $\mathbb{R}^n$ and let t be a positive number. Let Q_1, Q_2, ... be the collection of maximal dyadic cubes satisfying the condition*
$$|Q|^{-1} \int_Q |f|\, \mathrm{d}\mathcal{L}_n > t.$$

1. *Prove that there exists a constant $c > 0$ depending only on the dimension n such that, for all j and all $x \in Q_j$, one has $\mathsf{M}f(x) > ct$.*

2. *Prove that $\mathcal{L}_n(\{\mathsf{M}f > ct\}) \geq 2^{-n}t^{-1}\int_{|f|>t}|f|\,\mathrm{d}\mathcal{L}_n$.*

Exercise 3.2. *Let f be a nonzero function, integrable on $\mathbb{R}^n$ and vanishing outside a ball B. Prove that if $\mathsf{M}f$ is locally integrable, then $\int_{\mathsf{B}}|f|\log^+|f|\,\mathrm{d}\mathcal{L}$ is finite.*
Hint: *Use the preceding exercise.*

Exercise 3.3. *Show that the conclusions of the first theorem on singular integrals (Theorem 3.5) still hold if one replaces the hypothesis $\|\widehat{K}\|_\infty \leq B$ by "there exists $p_0 \in (1, +\infty)$ such that the operator T is bounded from L^{p_0} to L^{p_0} with norm $\leq B$".*

Exercise 3.4. *Let $K \in L^2(\mathbb{R}^n \times \mathbb{R}^n)$. For $f \in L^2(\mathbb{R}^n)$, set*

$$Tf(x) = \int_{\mathbb{R}^n} K(x, y)\, f(y)\,\mathrm{d}y \quad \text{for almost every } x.$$

Moreover, assume that

$$\sup_{y,\, z \in \mathbb{R}^n} \int_{|x-z| \geq 2|y-z|} |K(x, y) - K(x, z)|\,\mathrm{d}x \leq B < +\infty.$$

Show that there exists $C > 0$ depending only on n, $\|K\|_2$, and B such that, for any $f \in L^1(\mathbb{R}^n) \cap L^2(\mathbb{R}^n)$ and any $t > 0$, one has

$$\left|\{|Tf| > t\}\right| \leq \frac{C}{t}\,\|f\|_{L^1(\mathbb{R}^n)}.$$

Hint: *Use Exercise A.5 for the L^2-boundedness of T and use a of Calderón–Zygmund decomposition as in the proof of the first theorem on singular integrals (Theorem 3.5).*

Exercise 3.5 (singular integral on $\mathbb{Z}$). *If $a = \big(a(j)_{j \in \mathbb{Z}}\big)$ is a sequence which has only finitely many nonzero terms, set*

$$Ha(k) = \sum_{j \neq 0} \frac{a(k - j)}{j}.$$

1. *Prove the identity*

$$\sum_{j \neq 0} \frac{\sin jx}{j} = \frac{\pi - x}{2} \quad \text{if} \quad 0 < x < 2\pi.$$

2. *Prove that $\|Ha\|_{\ell^2} \leq \pi\,\|a\|_{\ell^2}$.*

3. *Prove that* $\displaystyle\sup_{\substack{m\neq 0}} \sum_{\substack{n\in\mathbb{Z} \\ |n|\leq 2|m|}} \left| \frac{1}{n-m} - \frac{1}{n} \right| < +\infty.$

4. *Adapt the proof of Theorem 3.5 to show that there exists a positive constant C such that, for all a and for all $t > 0$, we have*

$$\mathrm{card}\{|Ha| > t\} \leq \frac{C}{t}\, \|a\|_{\ell^1}.$$

5. *Prove that for all $1 < p < +\infty$ there exists C_p such that, for all a, $\|Ha\|_{\ell^p} \leq C_p\|a\|_{\ell^p}$.*

Exercise 3.6. *A function m on $\mathbb{R}^n$ is a p-multiplier if there exists a constant $C_p > 0$ such that, for all $f \in L^2 \cap L^p(\mathbb{R}^n)$, one has $\|Tf\|_p \leq C_p\|f\|_p$, where $\widehat{Tf} = \mathsf{m}\hat{f}$.*
  If φ is a C^1 function on $\mathbb{R}^2$ with compact support, set

$$T\varphi(x,y) = \mathrm{pv} \int_{-\infty}^{+\infty} \frac{\varphi(x-t,y)}{t}\, \mathrm{d}t.$$

1. *Show that, for all $p \in (1, +\infty)$, there exists $C_p > 0$ such that, for every function φ as above, one has*
$$\|T\varphi\|_{L^p(\mathbb{R}^2)} \leq C_p\, \|\varphi\|_{L^p(\mathbb{R}^2)}.$$

2. *Compute the Fourier transform $\displaystyle\iint_{\mathbb{R}^2} T\varphi(x,y)\, \mathrm{e}^{-2\mathrm{i}\pi(ux+vy)}\mathrm{d}x\, \mathrm{d}y$ in terms of the Fourier transform of φ, which is still as above.*

3. *Show that the indicator function of the set $\{(x,y) \ : \ x > 0 \text{ and } y > 0\}$ is a p-multiplier for any $p \in (1, +\infty)$.*

4. *Show that the indicator function of the interior of a convex polygon is a p-multiplier for any $p \in (1, +\infty)$.*

Chapter 4

The Littlewood–Paley theory

4.1 ▪ Vector-valued singular integrals

Let $(X, \mathcal{A}, \mu)$ be a measure space and let $\mathcal{H}$, $\mathcal{H}_1$, and $\mathcal{H}_2$ be separable Hilbert spaces. Inner products and norms in these spaces are denoted by $\langle \, , \, \rangle$ and $| \, |$, respectively. A function f from X to $\mathcal{H}$ is $\mathcal{A}$-measurable if $x \mapsto \langle f(x), v \rangle$ is measurable for all $v \in \mathcal{H}$. The space $L^p(\mu, \mathcal{H})$ is the set of measurable functions from X to $\mathcal{H}$ such that $\int_X |f|^p \, d\mu < +\infty$. When $p \geq 1$, endowed with the norm

$$\|f\|_{L^p(\mu, \mathcal{H})} = \left(\int_X |f|^p \, d\mu \right)^{\frac{1}{p}},$$

$L^p(\mu, \mathcal{H})$ is a Banach space. The space $L^\infty(\mu, \mathcal{H})$ is also defined as in the usual case of scalar functions. To be more precise, these spaces are not spaces of functions, but rather spaces of classes of functions modulo the equality μ-almost everywhere. When no reference is made to a measure μ, it is assumed that μ is the Lebesgue measure on $\mathbb{R}^n$. When no reference is made to a Hilbert space, it is assumed we are dealing with scalar functions. So, for instance, L^p stands for the functions from $\mathbb{R}$ to $\mathbb{C}$ whose pth power is integrable.

When $1 \leq p < +\infty$, the dual space to $L^p(\mu, \mathcal{H})$ is $L^{p'}(\mu, \mathcal{H})$, where $p^{-1} + p'^{-1} = 1$, and the duality form is

$$(f, g) \longmapsto \int_X \langle f(x), g(x) \rangle \, d\mu(x), \qquad f \in L^p(\mu, \mathcal{H}), \ g \in L^{p'}(\mu, \mathcal{H}).$$

For $f \in L^1(\mu, \mathcal{H})$, the map $v \mapsto \int_X \langle v, f(x) \rangle \, d\mu(x)$ is a continuous linear form on $\mathcal{H}$. So there exists a unique vector $I \in \mathcal{H}$ such that, for all $v \in \mathcal{H}$, one has

$$\langle I, v \rangle = \int_X \langle v, f(x) \rangle \, d\mu(x).$$

This vector is by definition the integral of f; we set

$$\int_X f(x) \, d\mu(x) = I.$$

We formally have the same inequalities as usual: $\left| \int_X f \, d\mu \right| \leq \int_X |f| \, d\mu$ and the Hölder and Minkowski inequalities.

The space $\mathcal{L}(\mathcal{H}_1, \mathcal{H}_2)$ of bounded linear operators from $\mathcal{H}_1$ to $\mathcal{H}_2$ endowed with the operator norm, again denoted by $| \ |$, is complete.

A function K from X to $\mathcal{L}(\mathcal{H}_1, \mathcal{H}_2)$ is measurable if, for any $v_1 \in \mathcal{H}_1$ and $v_2 \in \mathcal{H}_2$, the function $x \mapsto \langle K(x)v_1, v_2 \rangle$ is measurable.

If K is a function from X to $\mathcal{L}(\mathcal{H}_1, \mathcal{H}_2)$ and f is a function from X to $\mathcal{H}_1$, the integral $\int_X K(x) f(x) \, d\mu(x)$ exists, and then is an element of $\mathcal{H}_2$, under conditions analogous to the scalar case, for instance, when $K \in L^p(\mu, \mathcal{L}(\mathcal{H}_1, \mathcal{H}_2))$ and $f \in L^{p'}(\mu, \mathcal{H}_1)$.

So, in case $X = \mathbb{R}^n$ and μ is the Lebesgue measure, one can consider the convolution $K * f$. The result is a function from $\mathbb{R}^n$ to $\mathcal{H}_2$.

The Fourier transform of $f \in L^1(\mathbb{R}^n, \mathcal{H})$ is so defined:

$$\langle \widehat{f}(y), u \rangle = \int_{\mathbb{R}^n} \langle f(x), u \rangle e^{-2i\pi x \cdot y} \, dx \quad \text{for all} \quad u \in \mathcal{H}.$$

The Plancherel theorem holds for $L^2(\mathbb{R}^n, \mathcal{H})$.

Similarly the Fourier transform of $K \in L^1(\mathbb{R}^n, \mathcal{L}(\mathcal{H}_1, \mathcal{H}_2))$ is defined by the following equation:

$$\langle \widehat{K}(y)u, v \rangle = \int_{\mathbb{R}^n} \langle K(x)u, v \rangle e^{-2i\pi x \cdot y} \, dx \quad \text{for all} \quad u \in \mathcal{H}_1, \ v \in \mathcal{H}_2.$$

With this notation, the previous theorems concerning the singular integrals remain formally the same. These vector-valued versions are not a consequence of the scalar case, but their proofs follow verbatim the same lines.

4.2 ▪ The Littlewood–Paley inequalities

Let γ be a C^∞-function whose support is contained in the interval $[\alpha, \beta]$, with $0 < \alpha < \beta$. Consider the function φ from $\mathbb{R}^n$ to $\mathbb{C}$ such that $\widehat{\varphi}(y) = \gamma(|y|)$, and for $j \in \mathbb{Z}$ set $\varphi_j(x) = 2^{-nj} \varphi(2^{-j} x)$.

Since $y \mapsto \gamma(|y|)$ is in the Schwartz space $\mathscr{S}(\mathbb{R}^n)$, so is φ. This, in particular, implies that there exists $C > 0$ such that, for all x,

$$|\partial \varphi(x)| \leq C \min\{1, |x|^{-(n+2)}\} \leq C \min\{1, |x|^{-(n+1)}\}, \tag{4.1}$$

where $\partial \varphi$ stands for any first-order partial derivative of φ.

Theorem 4.1. *For any $p \in (1, +\infty)$ there exists $C_p > 0$ such that*

1. *for any sequence $(f_j)_{j \in \mathbb{Z}}$ of L^p-functions, one has*

$$\left\| \left(\sum |\varphi_j * f_j|^2 \right)^{\frac{1}{2}} \right\|_p \leq C_p \left\| \left(\sum |f_j|^2 \right)^{\frac{1}{2}} \right\|_p; \tag{4.2}$$

2. *for any $f \in L^p$, one has*

$$\left\| \left(\sum |\varphi_j * f|^2 \right)^{\frac{1}{2}} \right\|_p \leq C_p \|f\|_p; \tag{4.3}$$

3. *for any sequence* $(f_j)_{j\in\mathbb{Z}}$ *of* L^p-*functions such that* $\left\|\left(\sum |f_j|^2\right)^{\frac{1}{2}}\right\|_p < +\infty$, *the series* $\sum \varphi_j * f_j$ *converges in* L^p *and one has*

$$\left\|\sum \varphi_j * f_j\right\|_p \le C_p \left\|\left(\sum |f_j|^2\right)^{\frac{1}{2}}\right\|_p ; \tag{4.4}$$

4. *if, for all* $t > 0$, $\sum \gamma(2^j t)^2 = 1$, *or if, for all* $t > 0$, $\sum \gamma(2^j t) = 1$, *then, for all* $f \in L^p$, *one has*

$$\|f\|_p \le C_p \left\|\left(\sum |\varphi_j * f|^2\right)^{\frac{1}{2}}\right\|_p.$$

Proof of assertion 1.

The Hilbert space $\mathscr{H} = \ell^2(\mathbb{Z})$ has a natural orthonormal basis $(e_j)_{j\in\mathbb{Z}}$: all the coordinates of e_j are 0, except the jth, which is 1.

For $x \ne 0$ in $\mathbb{R}^n$, let $K(x)$ be the operator on $\mathscr{H}$ such that $K(x) e_j = \varphi_j(x) e_j$. We are going to show that the kernel K fulfills the hypotheses of Theorem 3.5.

We have $\widehat{K}(y) e_j = \gamma(2^j |y|) e_j$. This means that we have $\left\|\widehat{K}\right\|_\infty = \|\gamma\|_\infty$.

Now we estimate the gradient ∇K. Let ∂ stand for the partial derivation with respect to any of the components of $x = (x_1, \ldots, x_n)$. We have $\partial K(x) e_j = 2^{-(n+1)j} (\partial\varphi)(2^{-j} x) e_j$. This, together with (4.1), implies that for $x \ne 0$ we have

$$|\partial K(x)| \le C \sup_{j\in\mathbb{Z}} 2^{-(n+1)j} \min\{1, 2^{(n+1)j} |x|^{-(n+1)}\} \le \frac{C}{|x|^{n+1}}.$$

Therefore, due to Lemma 3.6 and the vector version of Theorem 3.5, the operator T associated with the kernel K is bounded on L^p for $1 < p < +\infty$, which means

$$\left\|\left(\sum_{j\in\mathbb{Z}} |\varphi_j * f_j|^2\right)^{1/2}\right\|_p = \|Tf\|_{L^p(\mathscr{H})}$$

$$\le C_p \|f\|_{L^p(\mathscr{H})} = C_p \left\|\left(\sum_{j\in\mathbb{Z}} |f_j|^2\right)^{1/2}\right\|_p.$$

Proof of assertion 2.

This time, let us set $K(x) = \sum_{j\in\mathbb{Z}} \varphi_j(x) e_j$. Since φ is rapidly decreasing this sum defines a function from $\mathbb{R}^n \setminus \{0\}$ to $\mathscr{H}$, or equivalently a function from $\mathbb{R}^n$ to $\mathcal{L}(\mathbb{C}, \mathscr{H})$.

We have $\widehat{K}(y) = \sum_{j\in\mathbb{Z}} \gamma(2^j |y|) e_j$. But for $y \ne 0$, there is at most $1 + \log(\beta/\alpha)/\log 2$ nonzero terms in this sum, since we must have $\alpha/|y| < 2^{-j} < \beta/|y|$. So $\left|\widehat{K}(y)\right|_{\mathscr{H}}^2 \le (1 + \log(\beta/\alpha)/\log 2) \|\gamma\|_\infty^2$.

We have, for $|x| \neq 0$,

$$
\begin{aligned}
|\partial K(x)|^2 &= \sum_{j \in \mathbb{Z}} 2^{-2(n+1)j} |(\partial \varphi)(2^{-j} x)|^2 \\
&\leq C \sum_{j \in \mathbb{Z}} 2^{-2(n+1)j} \min\{1, 2^{2(n+2)j} |x|^{-2(n+2)}\}.
\end{aligned}
$$

Let j_0 be the integer defined by the inequalities $2^{j_0} \leq |x| < 2^{j_0+1}$. We have

$$
\begin{aligned}
|\partial K(x)|^2 &\leq C \sum_{j > j_0} 2^{-2(n+1)j} + C \sum_{j \leq j_0} 2^{-2(n+1)j} 2^{2(n+2)j} |x|^{-2(n+2)} \\
&= C \sum_{j > j_0} 2^{-2(n+1)j} + C \sum_{j \leq j_0} 2^{2j} |x|^{-2(n+2)} \\
&\leq C' \left(2^{-2(n+1)j_0} + 2^{2j_0} |x|^{-2(n+2)} \right) \leq C'' |x|^{-2(n+1)}.
\end{aligned}
$$

Therefore, as previously mentioned, the operator T associated with the kernel K is bounded on L^p for $1 < p < +\infty$. If $f \in L^p$, one has

$$
Tf = \sum_{j \in \mathbb{Z}} \varphi_j * f \, e_j,
$$

so

$$
\left\| \left(\sum_{j \in \mathbb{Z}} |\varphi_j * f|^2 \right)^{1/2} \right\|_p = \|Tf\|_{L^p(\mathscr{H})} \leq C_p \|f\|_{L^p}.
$$

Proof of assertion 3.

Consider the operator T of the previous section. Let J be a finite subset of $\mathbb{Z}$. If $f = \sum_{j \in J} f_j e_j \in L^p(\mathscr{H})$ and $g \in L^{p'}$ with $p^{-1} + p'^{-1} = 1$, one has

$$
\begin{aligned}
\int_{\mathbb{R}^n} \langle f(x), Tg(x) \rangle_{\mathscr{H}} \, \mathrm{d}x &= \sum_{j \in J} \int_{\mathbb{R}^n} f_j(x) \overline{\varphi_j * g(x)} \, \mathrm{d}x \\
&= \sum_{j \in J} \int_{\mathbb{R}^n} f_j * \varphi_j(x) \overline{g(x)} \, \mathrm{d}x \\
&= \int_{\mathbb{R}^n} \left(\sum_{j \in J} f_j * \varphi_j(x) \right) \overline{g(x)} \, \mathrm{d}x.
\end{aligned}
$$

This means that the adjoint operator of T, considered as an operator from $L^{p'}$ to $L^{p'}(\mathscr{H}_N)$, sends $\sum_{j \in J} f_j e_j$ on $\sum_{j \in J} f_j * \varphi_j$. This proves that

$$
\left\| \sum_{j \in J} f_j * \varphi_j \right\|_p \leq C_p \left\| \left(\sum_{j \in J} |f_j|^2 \right)^{1/2} \right\|_p. \tag{4.5}
$$

Now, if $f = (f_j)_{j \in \mathbb{Z}}$ is a sequence of functions such that $\left\| \left(\sum_{|j| \in \mathbb{Z}} |f_j|^2 \right)^{1/2} \right\|_p$ is finite, (4.5) implies that $\sum_{|j| \leq N} f_j * \varphi_j$ is a Cauchy sequence in L^p, hence the conclusion.

Proof of assertion 4.

In (4.4) take $f_j = \varphi_j * f$. Then we have

$$\left\| \sum (\varphi_j * \varphi_j) * f \right\|_p \leq C_p \left\| \left(\sum |\varphi_j * f|^2 \right)^{1/2} \right\|_p.$$

But if $\sum \gamma(2^j t)^2 = 1$, we have $\sum \varphi_j * \varphi_j * f = f$ (this can be seen by taking the Fourier transform).

For dealing with the case $\sum \gamma(2^j t) = 1$, consider a C^∞-function $\tilde\varphi$ supported on an interval $[\tilde\alpha, \tilde\beta]$ with $0 < \tilde\alpha < \alpha < \beta < \tilde\beta$ and such that $\tilde\varphi$ is constant and equal to 1 on the interval $[\alpha, \beta]$. Define $\tilde\varphi_j$ as φ_j was defined, but using $\tilde\gamma$ instead of γ. By looking at Fourier transforms, we see that $\varphi_j * \tilde\varphi_j = \varphi_j$. So, if in (4.4) we replace φ_j by $\tilde\varphi_j$ and f_j by $\varphi_j * f$, we get

$$\|f\|_p = \left\| \sum \varphi_j * f \right\|_p$$
$$= \left\| \sum \tilde\varphi_j * (\varphi_j * f) \right\|_p$$
$$\leq C_p \left\| \left(\sum |\varphi_j * f|^2 \right)^{1/2} \right\|_p.$$

4.3 ▪ The Marcinkiewicz multiplier theorem

4.3.1 ▪ The 1-dimensional case

In this section $n = 1$ and H stands for the Hilbert transform.

Lemma 4.2. *Let $(X, \mathcal{A}, \mu)$ be a σ-finite measure space. For any $p \in (1, +\infty)$ there exists C_p such that, for all $f \in L^p(L^2(\mu))$, one has*

$$\left\| \left(\int |Hf_t|^2 d\mu(t) \right)^{1/2} \right\|_p \leq C_p \left\| \left(\int |f_t|^2 d\mu(t) \right)^{1/2} \right\|_p.$$

Proof. Let $L^p(L^2(\mu))$ denote the set of (classes of) measurable functions $(x, t) \in \mathbb{R} \times X \longmapsto f_t(x)$ such that

$$\left\| \left(\int_X |f_t|^2 d\mu(t) \right)^{1/2} \right\|_{L^p(\mathbb{R})} < +\infty.$$

For $x \neq 0$ in $\mathbb{R}$, let us consider the operator $K(x)$ of $L^2(\mu)$ which simply is the multiplication by $1/x$. It is clear that the norms of $K(x)$ and of its derivative are, respectively, bounded by $|x^{-1}|$ and x^{-2}. So K is a kernel which defines a bounded operator T on any space $L^p(L^2(\mu))$, with $1 < p < +\infty$, according to the vector version of Theorem 3.7. As we have $(Tf)_t(x) = H(f_t)(x)$, the conclusion of the lemma just expresses the boundedness of T.

If I is an interval, bounded or not, of $\mathbb{R}$, let S_I stand for the operator corresponding to the Fourier multiplier $\mathbf{1}_I$.

Corollary 4.3. *Let $(X, \mathcal{A}, \mu)$ be a measure space. For any $p \in (1, +\infty)$, there exists C_p such that, for any $\mathcal{A}$-measurable collection $\{I_t\}_{t \in X}$ of intervals and any $f \in L^p(L^2(\mu))$, we have*

$$\left\| \left(\int |S_{I_t} f_t|^2 \mathrm{d}\mu(t) \right)^{1/2} \right\|_p \le C_p \left\| \left(\int |f_t|^2 \mathrm{d}\mu(t) \right)^{1/2} \right\|_p.$$

Proof. This results from the fact, seen in Section 2.4.1, that the operator S_I is obtained by suitably composing the operator $\mathcal{H}$, isometries ϖ_t and the involution $\check{C}$.

Theorem 4.4. *Let $p \in (1, +\infty)$ and let $f \in L^p$. For $j \in \mathbb{Z}$, set*

$$\Delta_{j,+} = S_{[2^j, 2^{j+1}]} f \qquad and \qquad \Delta_{j,-} = S_{[-2^{j+1}, -2^j]} f.$$

Then, we have

$$C_p^{-1} \|f\|_p \le \left\| \left(\sum_{j \in \mathbb{Z}} |\Delta_{j,+}|^2 + |\Delta_{j,-}|^2 \right)^{1/2} \right\|_p \le C_p \|f\|_p,$$

where C_p does not depend on f.

Proof. Set $\chi = \mathbf{1}_{[1,2)}$, consider the function ψ such that $\widehat{\psi}(y) = \chi(|y|)$ and define $\psi_j(x) = 2^{-j}\psi(2^{-j}x)$. Choose for the function γ of Section 4.2 a C^∞-function vanishing outside the interval $[1/2, 4]$ and identically 1 on the interval $[1, 2]$.

We then have the following facts:

1. $\sum_{j \in \mathbb{Z}} \widehat{\psi_j} = 1$ almost everywhere.

2. $\varphi_j * \psi_j = \psi_j$.

3. $\psi_j * f = \Delta_{j,+} + \Delta_{j,-}$.

4. $\psi_j * \mathcal{H}f = \Delta_{j,+}$, where $\mathcal{H} = \frac{1}{2}\left(\mathrm{Id} + \frac{i}{\pi}\right)$ (see Section 2.4.1).

5. $\psi_j * (\mathrm{Id} - \mathcal{H})f = \Delta_{j,-}$.

By putting $f_j = \psi_j * f$ into (4.4) we obtain

$$\|f\|_p = \left\| \sum \psi_j * f \right\|_p = \left\| \sum \varphi_j * \psi_j * f \right\|_p$$

$$\le C_p \left\| \left(\sum |\psi_j * f|^2 \right)^{1/2} \right\|_p$$

$$\le C_p \left\| \left(2 \sum |\Delta_{j,+}|^2 + |\Delta_{j,-}|^2 \right)^{1/2} \right\|_p.$$

To get the converse inequality, we use Corollary 4.3 with the counting measure on $\mathbb{Z} \times \{+,-\}$, $I_{j,+} = [2^j, 2^{j+1}]$, $I_{j,-} = [-2^{j+1}, -2^j]$, and $f_{j,\pm} = \varphi_j * f$. We get

$$\left\| \left(\sum |\Delta_{j,+}|^2 + |\Delta_{j,-}|^2 \right)^{1/2} \right\|_p \leq C_p \left\| \left(2 \sum |\varphi_j * f|^2 \right)^{1/2} \right\|_p.$$

We conclude by using (4.3).

Theorem 4.5 (Marcinkiewicz multiplier theorem). *Let* m *be a bounded C^1-function from $\mathbb{R} \setminus \{0\}$ to $\mathbb{C}$ such that* $\displaystyle\sup_{j \in \mathbb{Z}} \int_{2^j \leq |x| \leq 2^{j+1}} |m'(x)| \, dx = B < +\infty$. *Then the operator defined by the Fourier multiplier* m *is bounded on $L^p(\mathbb{R})$ for any $p \in (1, +\infty)$.*

Proof. We have to prove that, for $f \in L^p \cap L^2$, if Tf is defined such that its Fourier–Plancherel transform is $m \widehat{f}$, we have $\|Tf\|_p \leq C_p \|f\|_p$ for a suitable C_p independent of f. Due to the boundedness of the Hilbert transform on L^p, we may separately consider the case when $\widehat{f}$ vanishes on $(-\infty, 0)$ and the case when $\widehat{f}$ vanishes on $(0, +\infty)$. Let us only treat the first case (the second one can be dealt with in the same way).

Let J_j stand for the interval $[2^j, 2^{j+1}]$. Due to Theorem 4.4, we have

$$\|Tf\|_p \leq C_p \left\| \left(\sum |S_{J_j} Tf|^2 \right)^{1/2} \right\|_p$$

and

$$\left\| \left(\sum |S_{J_j} f|^2 \right)^{1/2} \right\|_p \leq C_p \|f\|_p.$$

If $J = [a, b]$ is any one of the intervals J_j, we have $S_J Tf = S_J T(S_J f)$ and

$$\widehat{S_J Tf}(y) = m(y) \widehat{S_J f}(y) = \left(m(a) + \int_a^y m'(t) \, dt \right) \widehat{S_J f}(y) \tag{4.6}$$

$$= m(a) \widehat{S_J f}(y) + \int_a^b \mathbf{1}_{[t,b]}(y) \, \widehat{S_J f}(y) \, m'(t) \, dt \tag{4.7}$$

$$= m(a) \widehat{S_J f}(y) + \int_a^b \left(S_{[t,b]} S_J f \right)\widehat{}(y) \, m'(t) \, dt. \tag{4.8}$$

In other words, we have

$$S_J Tf = m(a) \, S_J f + \int_a^b S_{[t,b]}(S_J f) \, m'(t) \, dt.$$

The Cauchy–Schwarz inequality then yields

$$|S_J Tf|^2 \leq \left(|m(a)| + \int_a^b |m'(t)| \, dt \right) \left(|m(a)| \, |S_J f|^2 + \int_a^b |S_{[t,b]}(S_J f)|^2 \, |m'(t)| \, dt \right)$$

$$\leq (\|m\|_\infty + B) \left(|m(a)| \, |S_J f|^2 + \int_a^b |S_{[t,b]}(S_J f)|^2 \, |m'(t)| \, dt \right).$$

Then

$$
\left\| \left(\sum |S_{J_j} T f|^2 \right)^{1/2} \right\|_p \leq
$$
$$
C_p(\|\mathsf{m}\|_\infty + B)^{1/2} \left\| \left(\sum \left(|\mathsf{m}(2^j)| |S_{J_j} f|^2 + \int_{2^j}^{2^{j+1}} |S_{[t,2^{j+1}]} S_{J_j} f|^2 |\mathsf{m}'(t)| \, dt \right) \right)^{1/2} \right\|_p
$$

By using Corollary 4.3 with

- $\mu = \sum\limits_{j \in \mathbb{Z}} |\mathsf{m}(2^j)| \delta_{2^j} + |\mathsf{m}'(t)| \, dt$,

- $I_t = [t, 2^{j+1}]$ if $2^j \leq t < 2^{j+1}$,

- $f_t = S_{J_j} f$ if $2^j \leq t < 2^{j+1}$,

we get

$$
\left\| \left(\sum |S_{J_j} T f|^2 \right)^{1/2} \right\|_p
$$
$$
\leq C_p(\|\mathsf{m}\|_\infty + B)^{1/2} \left\| \left(\sum \left(|\mathsf{m}(2^j)| + \int_{2^j}^{2^{j+1}} |\mathsf{m}'(t)| \, dt \right) |S_{J_j} f|^2 \right)^{1/2} \right\|_p
$$
$$
\leq C_p(\|\mathsf{m}\|_\infty + B) \left\| \left(\sum |S_{J_j} f|^2 \right)^{1/2} \right\|_p.
$$

This means

$$
\left\| \left(\sum |\Delta_j T f|^2 \right)^{1/2} \right\|_p \leq C_p(\|\mathsf{m}\|_\infty + B) \left\| \left(\sum |\Delta_j f|^2 \right)^{1/2} \right\|_p.
$$

We conclude by using Theorem 4.4.

4.3.2 ▪ Higher dimensions

In order to keep notation simple we deal with dimension 2. Readers are invited to work the general case on their own.

We define two transformations,

$$
H_1 f(x,y) = \mathrm{pv} \int_{-\infty}^{+\infty} \frac{f(x - t, y)}{t} \, dt \quad \text{and} \quad H_2 f(x,y) = \mathrm{pv} \int_{-\infty}^{+\infty} \frac{f(x, y - t)}{t} \, dt,
$$

where, for the moment, $f \in \mathscr{S}(\mathbb{R}^2)$.

Since the Hilbert transform is bounded on $L^p(\mathbb{R})$ (for $1 < p < +\infty$), we have

$$
\int_{\mathbb{R}} |H_1 f(x,y)|^p dx \leq C_p^p \int_{\mathbb{R}} |f(x,y)|^p dx,
$$
$$
\iint_{\mathbb{R}^2} |H_1 f(x,y)|^p dx dy \leq C_p^p \iint_{\mathbb{R}^2} |f(x,y)|^p dx dy,
$$

and the same for H_2 (C_p is independent of f). This allows us to extend both H_1 and H_2 as bounded operators on $L^p(\mathbb{R}^2)$ (of course with $1 < p < +\infty$).

Lemma 4.6. *Let $(X\mathcal{A}, \mu)$ be a σ-finite measure space. For any $p \in (1, +\infty)$ there exists C_p such that, for all $f \in L^p(\mathbb{R}^2, L^2(\mu))$, one has*

$$\left\| \left(\int |H_i f_t|^2 d\mu(t) \right)^{1/2} \right\|_{L^p(\mathbb{R}^2)} \leq C_p \left\| \left(\int |f_t|^2 d\mu(t) \right)^{1/2} \right\|_{L^p(\mathbb{R}^2)},$$

for both $i = 1$ and $i = 2$.

Proof. This is an easy consequence of Lemma 4.2.

If I is a rectangle (i.e., a Cartesian product of intervals, bounded or not) let S_I stand for the operator associated with the multiplier $\mathbf{1}_I$.

Corollary 4.7. *Let $(X, \mathscr{A}, \mu)$ be a measure space. For any $p \in (1, +\infty)$, there exists C_p such that, for any $\mathcal{A}$-measurable collection $\{I_t\}_{t \in X}$ of rectangles and any $f \in L^p(\mathbb{R}^2, L^2(\mu))$, we have*

$$\left\| \left(\int |S_{I_t} f_t|^2 d\mu(t) \right)^{1/2} \right\|_{L^p(\mathbb{R}^2)} \leq C_p \left\| \left(\int |f_t|^2 d\mu(t) \right)^{1/2} \right\|_{L^p(\mathbb{R}^2)}.$$

Proof. This results from the fact that the operator S_I is obtained by suitably composing the operators $\mathcal{H}_1$, $\mathcal{H}_2$ (defined from H_1 and H_2, as $\mathcal{H}$ from H) and isometries.

Lemma 4.8. *Let $(X, \mathscr{A}, \mu)$ be a measure space. Let $(P_k)_{k \geq 1}$ be an orthogonal family of projectors of $L^2(\mu)$ such that $\sum P_k = \mathsf{Id}$. Assume that, for some $p \in (1, +\infty)$, there exists C such that, for all $f \in L^p(\mu) \cap L^2(\mu)$,*

$$\left\| \left(\sum_{k \geq 1} |P_k f|^2 \right)^{\frac{1}{2}} \right\|_{L^p(\mu)} \leq C_p \|f\|_{L^p(\mu)}.$$

Then, for all $f \in L^{p'}(\mu) \cap L^2(\mu)$,

$$\|f\|_{L^{p'}(\mu)} \leq C \left\| \left(\sum_{k \geq 1} |P_k f|^2 \right)^{\frac{1}{2}} \right\|_{L^{p'}(\mu)}.$$

Proof. For $f \in L^{p'}(\mu) \cap L^2(\mu)$ and $g \in L^p(\mu) \cap L^2(\mu)$, we have

$$\int f \bar{g} \, d\mu = \sum_{k \geq 1} \int P_k f \, \overline{P_k g} \, d\mu.$$

Let $g \in L^p(\mu) \cap L^2(\mu)$ be such that $g = \sum_{k=1}^{\nu} P_k g$ for some integer ν. We have

$$
\left| \int f\bar{g} \, \mathrm{d}\mu \right| = \left| \sum_{k=1}^{\nu} \int P_k f \, \overline{P_k g} \, \mathrm{d}\mu \right|
$$

$$
\leq \int \left| \sum_{k=1}^{\nu} P_k f \, \overline{P_k g} \right| \, \mathrm{d}\mu
$$

$$
\leq \int \left(\sum_{k=1}^{\nu} |P_k f|^2 \right)^{\frac{1}{2}} \left(\sum_{k=1}^{\nu} |P_k g|^2 \right)^{\frac{1}{2}} \, \mathrm{d}\mu
$$

$$
\leq \left\| \left(\sum_{k \geq 1} |P_k f|^2 \right)^{\frac{1}{2}} \right\|_{L^{p'}(\mu)} \left\| \left(\sum_{k \geq 1} |P_k g|^2 \right)^{\frac{1}{2}} \right\|_{L^{p}(\mu)}
$$

$$
\leq C \left\| \left(\sum_{k \geq 1} |P_k f|^2 \right)^{\frac{1}{2}} \right\|_{L^{p'}(\mu)} \|g\|_{L^p(\mu)}.
$$

Since the functions g are dense in $L^p(\mu)$, the lemma is proven.

Theorem 4.9. *Let $p \in (1, +\infty)$ and $f \in L^p(\mathbb{R}^2)$. For $j \in \mathbb{Z}$, set*

$$
\Delta_{j,k,\varepsilon,\eta} f = \Delta_{[\varepsilon 2^j, \varepsilon 2^{j+1}] \times [\eta 2^k, \eta 2^{k+1}]} f,
$$

for ε and η in $\{+, -\}$. Then we have

$$
C_p^{-1} \|f\|_{L^p(\mathbb{R}^2)} \leq \left\| \left(\sum_{\varepsilon, \eta \in \{+, -\}} \sum_{j, k \in \mathbb{Z}} |\Delta_{j,k,\varepsilon,\eta}|^2 \right)^{1/2} \right\|_{L^p(\mathbb{R}^2)} \leq C_p \|f\|_{L^p(\mathbb{R}^2)},
$$

where C_p does not depend on f.

Proof. The right-hand side inequality results from Corollary 4.7. The first inequality then comes from Lemma 4.8.

Theorem 4.10 (2-dimensional Marcinkiewicz multiplier theorem). *Let m be a bounded C^2-function from $\mathbb{R}^2 \setminus \{(0,0)\}$ to $\mathbb{C}$ such that*

1.

$$
\sup_{y \in \mathbb{R}, j \in \mathbb{Z}} \int_{2^j < |x| < 2^{j+1}} \left| \frac{\partial \mathrm{m}(x,y)}{\partial x} \right| \, \mathrm{d}x < +\infty;
$$

2.

$$
\sup_{x \in \mathbb{R}, k \in \mathbb{Z}} \int_{2^k < |y| < 2^{k+1}} \left| \frac{\partial \mathrm{m}(x,y)}{\partial y} \right| \, \mathrm{d}y < +\infty;
$$

3.

$$
\sup_{j,k \in \mathbb{Z}} \iint_{\substack{2^j < |x| < 2^{j+1} \\ 2^k < |y| < 2^{k+1}}} \left| \frac{\partial^2 \mathrm{m}(x,y)}{\partial x \partial y} \right| \, \mathrm{d}x\mathrm{d}y < +\infty.
$$

Then the operator defined by the Fourier multiplier m *is bounded on* $L^p(\mathbb{R}^2)$ *for any* $p \in (1, +\infty)$.

Proof. The main modification to the proof of Theorem 4.5 is to write

$$\mathsf{m}(x, y) = \iint_{[a,x] \times [b,y]} \frac{\partial^2 \mathsf{m}(u, v)}{\partial u \partial v} \, du dv + \mathsf{m}(x, b) + \mathsf{m}(a, y) - \mathsf{m}(a, b)$$

in formula (4.6).

4.4 ▪ Exercises

Exercise 4.1. *Let* m *be a bounded function from* $\mathbb{R} \setminus \{0\}$ *to* $\mathbb{C}$, *locally of bounded variation, and such that*

$$\sup_{j \in \mathbb{Z}} \int_{2^j \leq |t| \leq 2^{j+1}} |d\mathsf{m}|(t) < +\infty.$$

Show that, for all $p \in (1, +\infty)$, m *is a p-multiplier.*

Exercise 4.2. *Let* m *be a bounded* C^1-*function from* $\mathbb{R} \setminus \{0\}$ *to* $\mathbb{C}$ *such that* $\int_{-\infty}^{+\infty} |\mathsf{m}'(t)| \, dt$ $< \infty$ *and* $\lim_{t \to -\infty} \mathsf{m}(t) = 0$. *Show, without using the Marcinkiewicz multiplier theorem, that* m *is a multiplier for any* $p \in (1, +\infty)$.

   Hint: *For* $f \in L^2$ *define the operator* S_t, *so* $\widehat{S_t f} = \hat{f} \mathbf{1}_{(-\infty,t)}$, *and write* $\mathsf{m}(t) = \int_{-\infty}^t \mathsf{m}'(u) \, du$.

Exercise 4.3. *Let* m *be a bounded* C^1-*function from* $\mathbb{R} \setminus \{0\}$ *to* $\mathbb{C}$ *such that there exists a positive number* $\alpha > 1$ *such that*

$$\sup_{j \in \mathbb{Z}} \int_{\alpha^j}^{\alpha^{j+1}} |\mathsf{m}'(t)| \, dt < +\infty \qquad and \qquad \sup_{j \in \mathbb{Z}} \int_{-\alpha^{j+1}}^{-\alpha^j} |\mathsf{m}'(t)| \, dt < +\infty.$$

Show that m *is a multiplier for any* $p \in (1, +\infty)$.

Exercise 4.4. *Show that the function* $\mathsf{m}(x) = |x|^{\mathrm{i}}$ *defined on* $\mathbb{R} \setminus \{0\}$ *is a multiplier for every* $p \in (1, +\infty)$.

Exercise 4.5. *Let* m *be a* C^n-*function on* $\mathbb{R}^n \setminus \{0\}$ *such that, for every multi-index* α *with* $0 \leq |\alpha| \leq n$, *one has*

$$\sup_{x \in \mathbb{R}^n} |x|^{|\alpha|} |D^\alpha \mathsf{m}(x)| < +\infty.$$

Then, m *is a Fourier multiplier for all* $p \in (1, +\infty)$.

Remark. *This result is not accurate for $n \geq 3$; only $[n/2] + 1$ derivatives are needed (Mihlin theorem).*

Exercise 4.6. *Is $\left(1 + \left|\log|x|\right|\right)^{-1/2}$ a p-multiplier?*

Exercise 4.7. *A function $m \in L^\infty(\mathbb{R})$ is a p-multiplier if there exists a constant C_p such that, for all $f \in L^2(\mathbb{R}) \cap L^p(\mathbb{R})$, one has*

$$\|Tf\|_p \leq C_p\|f\|_p, \quad \text{where} \quad \widehat{Tf} = m \cdot \widehat{f}.$$

Let P be a polynomial with complex coefficients, of degree greater than or equal to 1, without real root.

1. *Prove that $\sup\limits_{x \in \mathbb{R}} 1/|P(x)|$ and $\sup\limits_{x \in \mathbb{R}} |P'(x)/P(x)|$ are finite.*

2. *Prove that, for all $p > 1$, $1/P(\log|x|)$ is a p-multiplier.*

Exercise 4.8. *Let H stand for the Hilbert space $L^2((0, +\infty), \frac{dx}{x})$. In other words, $g = (g_u)_{u>0}$ is in H if and only if $\int_0^{+\infty} |g_t|^2 \frac{dt}{t} < +\infty$, and its norm is $|g|_H = \left(\int_0^{+\infty} |g_t|^2 \frac{dt}{t}\right)^{\frac{1}{2}}$.*

Let χ be a C^∞-function from $(0, +\infty)$ to $\mathbb{R}$ whose support is included in an interval $[\alpha, \beta]$ with $0 < \alpha < \beta < \infty$. Consider the function φ defined on $\mathbb{R}^n$ such that $\widehat{\varphi}(y) = \chi(|y|)$ and set, for $u > 0$, $\varphi_u(y) = u^{-n}\varphi(u^{-1}y)$.

1. *Show that for all $p \in (1, +\infty)$ there exists a number C_p such that, for all $f \in L^p(\mathbb{R}^n, H)$ (written $(x, t) \mapsto f_t(x)$), one has*

$$\left\|\left(\int_0^{+\infty} |\varphi_f * f_t(x)|^2 \frac{dt}{t}\right)^{\frac{1}{2}}\right\|_{L^p(\mathbb{R})} \leq C_p \left\|\left(\int_0^{+\infty} |f_t(x)|^2 \frac{dt}{t}\right)^{\frac{1}{2}}\right\|_{L^p(\mathbb{R})}.$$

2. *Now suppose that $\int_0^{+\infty} |\chi(ut)|^2 \frac{dt}{t} = 1$ for all t. Give a sketch of the proof that, for all $p \in (1, +\infty)$, there exists C_p such that, for all $f \in L^p(\mathbb{R}^n)$, one has*

$$C_p^{-1}\|f\|_{L^p(\mathbb{R})} \leq \left\|\left(\int_0^{+\infty} |\varphi_f * f(x)|^2 \frac{dt}{t}\right)^{\frac{1}{2}}\right\|_{L^p(\mathbb{R})} \leq C_p \|f\|_{L^p(\mathbb{R})}.$$

Chapter 5

Higher Riesz transforms

5.1 ▪ Spherical harmonics

First we consider the set $\mathscr{P}_n$ of polynomials in n variables with complex coefficients as a subspace of $L^2(S_{n-1})$. So, if P and Q are two such polynomials, we set $(P, Q) = \int_{S_{n-1}} P\overline{Q} \, d\sigma$, where σ is the area measure on the unit sphere S_{n-1} of $\mathbb{R}^n$.

Let $\mathscr{P}_{n,k}$ stand for the subset of $\mathscr{P}_n$ of homogeneous polynomials of degree k. Let $\mathscr{H}_{n,k}$ stand for the set of harmonic polynomials in n variables which are homogeneous of degree k.

Lemma 5.1. *The spaces $\mathscr{H}_{n,k}$, for $k = 0, 1, \ldots$, are orthogonal.*

Proof. We use the Green formula: let P and Q be two homogeneous polynomials of respective degrees m and p, with $m \neq p$; we have

$$\int_{S_{n-1}} \left(P(x)\frac{\partial \overline{Q}(x)}{\partial n} - \overline{Q}(x)\frac{\partial P(x)}{\partial n} \right) d\sigma(x)$$
$$= \int_{|x|<1} \left(P(x)\Delta\overline{Q}(x) - \overline{Q}(x)\Delta P(x) \right) dx = 0.$$

As usual $\partial/\partial n$ stands for the outer normal derivative. On the unit sphere, $\frac{\partial P(x)}{\partial n} = x.\nabla P$; however, P being homogeneous of degree m, $x.\nabla P = mP(x)$. In the same way $\frac{\partial Q(x)}{\partial n} = pQ(x)$. It results that $(P, Q) = 0$.

In particular, any harmonic polynomial, homogeneous of positive degree k, is orthogonal to constants and therefore is of mean value 0 on the unit sphere. So, for such a P the kernel $\dfrac{P(x)}{|x|^{n+k}}$ is of the type considered in Theorem 3.8 on Calderón–Zygmund operators. So, there is an associated multiplier which we are going to compute.

We define on $\mathscr{P}_{n,k}$ a sesquilinear form: $\langle P, Q \rangle = P(D)\overline{Q}$. It is a scalar product; indeed the monomials form an orthogonal basis and

$$\left\langle \prod_{j=1}^n x_j^{\alpha_j}, \prod_{j=1}^n x_j^{\alpha_j} \right\rangle = \prod_{j=1}^n (\alpha_j!).$$

51

Lemma 5.2. *Let $P \in \mathscr{P}_{n,k}$, with $k \geq 2$. Then P can be written as $P(x) = Q(x) + |x|^2 R(x)$, where $Q \in \mathscr{H}_{n,k}$ and $R \in \mathscr{P}_{n,k-2}$.*

Proof. Let us determine the orthogonal space within $\mathscr{P}_{n,k}$ of $|x|^2 \mathscr{P}_{n,k-2}$ with respect to this last scalar product. If $P \in \mathscr{P}_{n,k}$ and $Q \in \mathscr{P}_{n,k-2}$, we have $\langle |x|^2 Q, P \rangle = \langle Q, \Delta P \rangle$. Therefore the orthogonal of $|x|^2 \mathscr{P}_{n,k-2}$ is the space $\mathscr{H}_{n,k}$. Hence the lemma.

Let $H_{n,k}$ stand for the space of restrictions to the unit sphere of elements of $\mathscr{H}_{n,k}$. The elements of $H_{n,k}$ are called the *spherical harmonic of degree k*, and those of $\mathscr{H}_{n,k}$ are referred to as the *solid spherical harmonics of degree k*.

A priori a spherical harmonic is only defined on the sphere, but it is convenient to extend it to $\mathbb{R}^n \setminus \{0\}$ as a 0-homogeneous function. So, if P is a solid spherical harmonic of degree k, the corresponding spherical harmonic Y is defined as $|x|^{-k} P(x)$ for all $x \neq 0$.

We say that a spherical harmonic is normalized if it has norm 1 in the space $L^2(S_{n-1})$. Lemma 5.2 has the following corollary.

Corollary 5.3.

1. *The restriction to the unit sphere of any polynomial is a sum of spherical harmonics.*

2. *The sums of spherical harmonics are dense in the space $C(S_{n-1})$.*

3. *Any $f \in L^2(S_{n-1})$ is the sum, convergent in $L^2(S_{n-1})$, of a unique series*

$$f = \sum_{k \geq 0} Y_k,$$

   where Y_k is a spherical harmonic of degree k. These terms are mutually orthogonal and

$$\|f\|_2^2 = \sum_{k \geq 0} \|Y_k\|_2^2.$$

Proof. The first assertion results from repeated use of Lemma 5.2. The second one then results from the Stone–Weierstrass theorem: the restriction of polynomials are dense in $C(S_{n-1})$. The third one comes from Lemma 5.1.

Lemma 5.4. *For every derivation multi-index α, there exists $c_{n,\alpha}$ such that, for all normalized spherical harmonic Y of degree $k \geq 1$, one has*

$$\sup_{x \in S_{n-1}} |D^\alpha Y(x)| \leq c_{n,\alpha} k^{n/2 + |\alpha|}.$$

Proof. Recall first that spherical harmonics are not only defined on the unit sphere but also are C^∞-functions on $\mathbb{R}^n \setminus \{0\}$. Let Y_k be a spherical harmonic of degree k. It is the restriction to the unit sphere of a harmonic polynomial P_k homogeneous of degree k: $Y_k(x) = |x|^{-k} P_k(x)$. Let φ be a nonnegative radial function supported in the unit ball and of integral 1. Due to the mean value property of harmonic functions, one has

$$P_k(x) = k^n \int P_k(y) \, \varphi\big(k(x - y)\big) \, \mathrm{d}y.$$

It results that, for every derivation multi-index α, one has

$$D^\alpha P_k(x) = k^{n+|\alpha|} \int_{|y| \leq 1/k} P_k(x-y) \, D^\alpha \varphi(ky) \, dy.$$

Then, if $|x| = 1$, one has

$$|D^\alpha P_k(x)|^2 \leq k^{2(n+|\alpha|)} \left(\int |D^\alpha \varphi(ky)|^2 dy \right) \left(\int_{\left||y|-1\right| \leq 1/k} |P_k(y)|^2 dy \right)$$

$$= k^{n+2|\alpha|} \|D^a \varphi\|^2_{L^2(\mathbb{R}^n)} \left(\int_{1-1/k}^{1+1/k} r^{n-1} \left(\int_{S_{n-1}} |P_k(ry)|^2 d\sigma(y) \right) dr \right)$$

$$\leq k^{n+2|\alpha|} \|D^a \varphi\|^2_{L^2(\mathbb{R}^n)} \|Y_k\|^2_{L^2(S_{n-1})} \int_0^{1+1/k} r^{n+2k-1} dr$$

$$\leq 2^n \|D^a \varphi\|^2_{L^2(\mathbb{R}^n)} k^{n+2|\alpha|} \left(1 + \frac{1}{k} \right)^{2k}$$

$$\leq 2^n e^2 \|D^a \varphi\|^2_{L^2(\mathbb{R}^n)} k^{n+2|\alpha|}.$$

The conclusion is then easily obtained by using the Leibniz formula for derivatives of a product.

Corollary 5.5. *If $f \in L^2(S_{n-1})$ has a development of the form*

$$f = \sum_{k \geq 0} a_k Y_k, \tag{5.1}$$

where Y_k is a normalized spherical harmonic of degree k and $|a_k| = O(k^{-m})$ for all $m \geq 1$, then f coincides almost everywhere with a C^∞-function.

Proof. By Lemma 5.4, for any α the series $\sum a_k D^\alpha Y_k$ converges uniformly.

To prove the converse of this corollary, we first need the following definition and lemma.

Definition 5.6. *A function f on the unit sphere is said to be in $C^k(S_{n-1})$ if the function $\widetilde{f} : x \longmapsto f(x/|x|)$ is C^k on $\mathbb{R}^n \setminus \{0\}$. For $f \in C^2(S_{n-1})$, one considers $\Delta_S f$ the restriction of $\Delta \widetilde{f}$ to S_{n-1} and calls Δ_S the spherical Laplacian.*

Lemma 5.7. *Let u and v be two C^2-functions on S_{n-1}. One has*

$$\int_{S_{n-1}} u \, \Delta_S v \, d\sigma = \int_{S_{n-1}} v \, \Delta_S u \, d\sigma.$$

Proof. Let $\widetilde{u}$ and $\widetilde{v}$ be the extensions of u and v as homogeneous functions of degree 0. As the derivative of a 0-homogeneous function along the normal to a sphere centered at 0 is zero, the Green formula applied to the annulus $\{x \in \mathbb{R}^n : 1 - \varepsilon \leq |x| \leq 1\}$ yields

$$\int_{1-\varepsilon \leq |x| \leq 1} \left(\widetilde{u}(x) \, \Delta \widetilde{v}(x) - \widetilde{v}(x) \, \Delta \widetilde{u}(x) \right) dx = 0.$$

However, one has $\Delta\widetilde{u}(x) = |x|^{-2}\Delta_S u(x/|x|)$, hence the result.

Lemma 5.8. *If Y is a spherical harmonic of degree k, one has*

$$\Delta_S Y = -k(n+k-2)\,Y.$$

Proof. One has $Y(x) = |x|^{-k}P(x)$, where P is a harmonic polynomial homogeneous of degree k. According to Definition 5.6, we have to compute $\Delta\big(|x|^{-k}P(x)\big)$ and take its restriction to the sphere.

One has $\Delta\big(|x|^{-k}P(x)\big) = |x|^{-k}\Delta P(x) + P(x)\,\Delta|x|^{-k} + 2(\nabla|x|^{-k}).\nabla P(x)$. But $\Delta P = 0$, $\nabla|x|^{-k} = -k\,|x|^{-(k+2)}x$, $\big(\nabla P(x)\big).x = k\,P(x)$, and $\Delta|x|^{-k} = -k(n-k-2)|x|^{-(k+2)}$. So

$$\Delta\big(|x|^{-k}P(x)\big) = -k(n+k-2)|x|^{-(k+2)}P(x).$$

Theorem 5.9. *A function $f \in L^2(S_{n-1})$ coincides almost everywhere with a C^∞-function if and only if the coefficients of its development* (5.1) *satisfy $|a_k| = \mathrm{O}(k^{-m})$ for all $m \geq 1$.*

Proof. In view of Corollary 5.5, it is enough to consider a C^∞-function f on the sphere. Consider its development (5.1).

One has $a_k = \displaystyle\int_{S_{n-1}} f(x)\overline{Y}_k(x)\,\mathrm{d}\sigma(x)$. Lemmas 5.7 and 5.8 give

$$\int_{S_{n-1}} \Delta_S f(x)\,\overline{Y}_k(x)\,\mathrm{d}\sigma(x) = \int_{S_{n-1}} f(x)\,\Delta_S\overline{Y}_k(x)\,\mathrm{d}\sigma(x)$$

$$= -k(n+k-2)\int_{S_{n-1}} f\,\overline{Y}_k\,\mathrm{d}\sigma.$$

By iterating this last formula and using the Cauchy–Schwarz inequality one gets

$$|a_k| \leq \big(k(n+k-2)\big)^{-m}\|\Delta_S^m f\|_{L^2(S_{n-1})}.$$

Later on, we shall need the following lemma.

Lemma 5.10. *Let Y be a spherical harmonic of degree $k > 0$. Then $x_1 Y(x) = Y_-(x) + Y_+(x)$, where Y_- and Y_+ are spherical harmonics of degrees $k-1$ and $k+1$, respectively.*

Proof. Let W be a spherical harmonic of degree $l \neq k \pm 1$. Consider the solid spherical harmonics $\widetilde{Y}(x) = |x|^k Y(x/|x|)$ and $\widetilde{W}(x) = |x|^l W(x/|x|)$ and set $u(x) = x_1\widetilde{Y}$. One has

$$\Delta u(x) = 2\,\frac{\partial\widetilde{Y}}{\partial x_1}.$$

By using the Green formula, one gets

$$(l - k - 1) \int_{S_{n-1}} x_1 Y(x) W(x) \, d\sigma(x) = \int_{S_{n-1}} \left(u(x) \frac{\partial \widetilde{W}(x)}{\partial n} - \widetilde{W}(x) \frac{\partial u(x)}{\partial n} \right) d\sigma(x)$$

$$= \int_{|x| \leq 1} \left(u(x) \Delta \widetilde{W}(x) - \widetilde{W}(x) \Delta u(x) \right) dx$$

$$= -2 \int_{|x| \leq 1} \widetilde{W} \frac{\partial \widetilde{Y}}{\partial x_1} \, dx = 0.$$

So, $x_1 Y$ is orthogonal to spherical harmonics of degree different from $k \pm 1$.

5.2 ▪ Higher Riesz transforms

Lemma 5.11 (Hecke identity). *Let P be a polynomial on $\mathbb{R}^n$, harmonic and homogeneous of degree $k \geq 0$. Then the Fourier transform of $P(x) \, \mathrm{e}^{-\pi |x|^2}$ is given by the following formula:*

$$\int_{\mathbb{R}^n} P(x) \, \mathrm{e}^{-\pi |x|^2} \mathrm{e}^{-2\mathrm{i}\pi x \cdot y} dx = (-\mathrm{i})^k P(y) \, \mathrm{e}^{-\pi |y|^2}. \tag{5.2}$$

Proof. We have, for real y,

$$\int_{\mathbb{R}^n} P(x) \, \mathrm{e}^{-\pi (|x|^2 - 2x \cdot y)} dx = \mathrm{e}^{\pi |y|^2} \int_{\mathbb{R}^n} P(x) \, \mathrm{e}^{-\pi |x-y|^2} dx$$

$$= \mathrm{e}^{\pi |y|^2} \int_{\mathbb{R}^n} P(x + y) \, \mathrm{e}^{-\pi |x|^2} dx$$

$$= P(y) \, \mathrm{e}^{\pi |y|^2}, \tag{5.3}$$

where the last equality comes from the mean property satisfied by the harmonic function P.

But both sides of (5.3) are analytic functions of y, so by replacing y with $-\mathrm{i}y$ in (5.3) one gets formula (5.2). ∎

Theorem 5.12. *Let P be a polynomial on $\mathbb{R}^n$, harmonic and homogeneous of degree $k > 0$. Then the multiplier associated with the singular convolution kernel* pv $\dfrac{P(x)}{|x|^{k+n}}$ *is the function* $\gamma_{n,k} \dfrac{P(y)}{|y|^k}$, *where* $\gamma_{n,k} = (-\mathrm{i})^k \pi^{n/2} \dfrac{\Gamma(k/2)}{\Gamma((k+n)/2)}$.

Proof. Recall that if f and g are integrable, one has

$$\int_{\mathbb{R}^n} f(x) \widehat{g}(x) \, dx = \int_{\mathbb{R}^n} \widehat{f}(y) g(y) \, dy. \tag{5.4}$$

By taking $f(x) = t^{\frac{n}{2}} P(t^{\frac{1}{2}} x) \, \mathrm{e}^{-\pi t |x|^2}$ (with $t > 0$ and P a harmonic polynomial of degree $k \geq 0$, later on k will be restricted to be positive) and g an integrable function, (5.2) yields, for any $t > 0$,

$$\int_{\mathbb{R}^n} t^{\frac{k+n}{2}} P(x) \, \mathrm{e}^{-\pi t |x|^2} \widehat{g}(x) \, dx = (-\mathrm{i})^k \int_{\mathbb{R}^n} t^{-\frac{k}{2}} P(y) \, \mathrm{e}^{-\pi |y|^2/t} g(y) \, dy.$$

Multiplying both sides by $t^{-(1+\alpha/2)}$, with $0 < \alpha < n$, and integrating with respect to t from 0 to $+\infty$ yields

$$\int_{\mathbb{R}^n} \frac{P(x)}{|x|^{n+k-\alpha}} \widehat{g}(x)\, \mathrm{d}x = (-\mathrm{i})^k \pi^{\frac{n}{2}-\alpha} \frac{\Gamma\left(\frac{k+\alpha}{2}\right)}{\Gamma\left(\frac{n+k-\alpha}{2}\right)} \int_{\mathbb{R}^n} \frac{P(y)}{|y|^{k+\alpha}} g(y)\, \mathrm{d}y. \qquad (5.5)$$

Now, we assume that P has a positive degree and $g \in \mathscr{S}(\mathbb{R}^n)$. Due to the regularity of g and due to the fact that P has mean 0 on the spheres, we can take the limits when α goes to 0 to obtain

$$\mathrm{pv} \int_{\mathbb{R}^n} \frac{P(x)}{|x|^{n+k}} \widehat{g}(x)\, \mathrm{d}x = (-\mathrm{i})^k \frac{\pi^{\frac{n}{2}} \Gamma\left(\frac{k}{2}\right)}{\Gamma\left(\frac{n+k}{2}\right)} \int_{\mathbb{R}^n} \frac{P(y)}{|y|^k} g(y)\, \mathrm{d}y. \qquad (5.6)$$

Taking limits deserves some explanations, at least for the left-hand side. We write

$$\int_{\mathbb{R}^n} \frac{P(x)}{|x|^{n+k-\alpha}} \widehat{g}(x)\, \mathrm{d}x = \int_{|x|<1} \frac{P(x)}{|x|^{n+k-\alpha}} \widehat{g}(x)\, \mathrm{d}x + \int_{|x|\geq 1} \frac{P(x)}{|x|^{n+k-\alpha}} \widehat{g}(x)\, \mathrm{d}x.$$

There is no problem with the second term. As we have

$$\int_{|x|<1} \frac{P(x)}{|x|^{n+k-\alpha}} \widehat{g}(x)\, \mathrm{d}x = \int_{|x|<1} \frac{P(x)}{|x|^{n+k-\alpha}} \left(\widehat{g}(x) - \widehat{g}(0)\right) \mathrm{d}x$$

we see that this term has the limit

$$\int_{|x|<1} \frac{P(x)}{|x|^{n+k}} \left(\widehat{g}(x) - \widehat{g}(0)\right) \mathrm{d}x,$$

which we can interpret as a principal value:

$$\int_{|x|<1} \frac{P(x)}{|x|^{n+k}} \left(\widehat{g}(x) - \widehat{g}(0)\right) \mathrm{d}x = \lim_{x \searrow 0} \int_{\varepsilon<|x|<1} \frac{P(x)}{|x|^{n+k}} \widehat{g}(x)\, \mathrm{d}x.$$

So, in (5.6) the first term should be interpreted as a principal value. We just proved that the Fourier transform as a distribution of $\mathrm{pv}\dfrac{P(x)}{|x|^{n+k}}$ is the function $\gamma_{n,k}\dfrac{P(y)}{|y|^k}$

Remark 5.13. *Another consequence of (5.5) is that the Fourier transform of $|x|^{\alpha-n}$ ($x \in \mathbb{R}^n$) is*

$$\pi^{\frac{n}{2}} \frac{\Gamma\left(\frac{\alpha}{2}\right)}{\Gamma\left(\frac{n-\alpha}{2}\right)} |y|^{-\alpha}$$

for $0 < \alpha < n$.

Corollary 5.14. *Let Ω be a 0-homogeneous function on $\mathbb{R}^n \setminus \{0\}$, such that*

$$\int_{S_{n-1}} |\Omega|^2\, \mathrm{d}\sigma < \infty \qquad and \qquad \int_{S_{n-1}} \Omega\, \mathrm{d}\sigma = 0.$$

Consider its expansion,

$$\Omega = \sum_{k\geq 1} Y_k, \ where\ Y_k\ is\ a\ spherical\ harmonic\ of\ degree\ k.$$

Then the multiplier associated with the singular convolution kernel is the function

$$\mathsf{m} = \sum_{k \geq 1} \gamma_{n,k} Y_k.$$

Proof. Let $\Omega_\nu = \sum_{k=1}^{\nu} Y_k$. Due to Theorem 5.12, the multiplier associated with the kernel $\Omega_\nu(x)|x|^{-n}$ is the function $\mathsf{m}_\nu = \sum_{k=1}^{\nu} \gamma_{n,k} Y_k$. But recall that this multiplier, for $|y| = 1$, is also expressed by the formula (see Theorem 2.8)

$$\mathsf{m}_\nu(y) = \int_{S_{n-1}} \Omega_\nu(x) \left(-i\frac{\pi}{2} \operatorname{sgn}(x \cdot y) + \log \frac{1}{|x \cdot y|} \right) d\sigma(x).$$

Since Ω_ν converges towards Ω in $L^2(S_{n-1})$ and the function $-i\frac{\pi}{2} \operatorname{sgn}(x \cdot y) + \log \frac{1}{x \cdot y}$ is square integrable on S_{n-1}, it results that m_ν converges uniformly on S_{n-1} towards

$$\mathsf{m}(y) = \int_{S_{n-1}} \Omega(x) \left(-i\frac{\pi}{2} \operatorname{sgn}(x \cdot y) + \log \frac{1}{|x \cdot y|} \right) d\sigma(x).$$

This ends the proof.

Theorem 5.15. *Let Ω be a C^∞ and 0-homogeneous function on $\mathbb{R}^n \setminus \{0\}$ of mean 0 on S_{n-1}. The multiplier associated with the singular integral kernel $\Omega(x)/|x|^n$ is a C^∞ and 0-homogeneous function m on $\mathbb{R}^n \setminus \{0\}$ of mean 0 on S_{n-1}. Conversely, such an m is the multiplier associated with one of the above kernels.*

Proof. Let us consider the development (5.1) of Ω. This development has no constant term since Ω has mean value 0. Due to Corollary 5.14 we know that the corresponding multiplier m is the sum of the uniformly convergent series

$$\sum_{j=1}^{k} \gamma_{n,j} a_j Y_j(y).$$

Due to Theorem 5.9 we know that $|a_\nu| = O(\nu^{-m})$ for all $m \geq 1$. So, it is the same for the coefficients $\gamma_{n,j} a_j$. Then Lemma 5.4 and Corollary 5.5 show that the sum of this series is C^∞.

Now, consider a C^∞ 0-homogeneous function m of mean 0 on the sphere. Consider its development,

$$\mathsf{m}(y) = \sum_{j \geq 1} a_j Y_j,$$

and the function $\Omega(x) = \sum_{j \geq 1} \gamma_{n,j}^{-1} a_j Y_j(x)$. This last series converges towards a C^∞-function because of the slow increase of $\gamma_{n,j}^{-1}$. Then the first part of the proof shows that m is the multiplier associated with the kernel $|x|^{-n}\Omega(x)$.

5.3 • Nonsmooth kernels

5.3.1 • Odd kernels

Theorem 5.16. *Let Ω be a 0-homogeneous odd function on $\mathbb{R}^n$, integrable with respect to the area measure on the sphere S_{n-1}. For $0 < \varepsilon < R$, let $T_{\varepsilon,R}$ be the convolution operator $f \longmapsto \left(x \mapsto \int_{\varepsilon < |y| < R} f(x-y) \dfrac{\Omega(y)}{|y|^n} \, \mathrm{d}y \right)$, and $T^* f = \sup\limits_{0 < \varepsilon < R} |T_{\varepsilon,R} f|$. Then, for all $p \in (1, +\infty)$,*

1. *there exists $C_p > 0$ such that $\|T^* f\|_p \le C_p \|f\|_p$ for all f;*

2. *for all $f \in L^p$, the limit $\lim\limits_{\substack{R \to \infty \\ \varepsilon \to 0}} T_{\varepsilon,R}$ exists and define a bounded operator T on L^p;*

3. *for all $f \in L^p$, $T_{\varepsilon,R} f$ converges almost everywhere to Tf as ε goes to 0 and R to $+\infty$.*

Proof. One has

$$
\begin{aligned}
T_{\varepsilon,R} f(x) &= \int_{\varepsilon < |y| < R} f(x-y) \frac{\Omega(y)}{|y|^n} \, \mathrm{d}y \\
&= \int_{S_{n-1}} \Omega(y) \left(\int_\varepsilon^R f(x-ry) \frac{\mathrm{d}r}{r} \right) \mathrm{d}\sigma(y) \\
&= \int_{S_{n-1}} \Omega(-y) \left(\int_\varepsilon^R f(x+ry) \frac{\mathrm{d}r}{r} \right) \mathrm{d}\sigma(y) \\
&= \int_{S_{n-1}} \Omega(y) \left(-\int_\varepsilon^R f(x+ry) \frac{\mathrm{d}r}{r} \right) \mathrm{d}\sigma(y) \\
&= \int_{S_{n-1}} \Omega(y) \left(\int_{-R}^{-\varepsilon} f(x-ry) \frac{\mathrm{d}r}{r} \right) \mathrm{d}\sigma(y),
\end{aligned}
$$

and therefore

$$
T_{\varepsilon,R} f(x) = \frac{1}{2} \int_{S_{n-1}} \Omega(y) \left(\int_{\varepsilon < |t| < R} f(x-ty) \frac{\mathrm{d}t}{t} \right) \mathrm{d}\sigma(y). \tag{5.7}
$$

Given $y \in S_{n-1}$ define the Hilbert transform H_y in direction y as the singular integral $H_y f(x) = \mathrm{pv} \int_{-\infty}^{+\infty} f(x-ty) \dfrac{\mathrm{d}t}{t}$ and the corresponding maximal operator $H_y^* f(x) = \sup_{\varepsilon > 0} \left| \int_{|t| > \varepsilon} f(x-ty) \frac{\mathrm{d}t}{t} \right|$.

Then it results from (5.7) that we have

$$
T^* f(x) \le \int_{S_{n-1}} |\Omega(y)| \, H_y^* f(x) \, \mathrm{d}\sigma(y).
$$

Then the integral Minkowski inequality writes as

$$
\|T^* f\|_p \le \int_{S_{n-1}} |\Omega(y)| \left(\int_{\mathbb{R}^n} (H_y^* f(x))^p \, \mathrm{d}x \right)^{1/p} \mathrm{d}\sigma(y). \tag{5.8}
$$

However, due to Theorem 3.12, for all $y \in S_{n-1}$ and for all $z \in y^{\perp}$, we have

$$\int_{\mathbb{R}} (H_y^* f(z + ty))^p \, \mathrm{d}t \leq C_p \int_{\mathbb{R}} |f(z + ty)|^p \, \mathrm{d}t$$

for some constant C_p depending only on p.

Therefore we have

$$\int_{\mathbb{R}^n} (H_y^* f(x))^p \, \mathrm{d}x = \int_{y^{\perp}} \left(\int_{\mathbb{R}} (H_y^* f(z + ty))^p \, \mathrm{d}t \right) \mathrm{d}z$$
$$\leq C_p \int_{y^{\perp}} \left(\int_{\mathbb{R}} |f(z + ty)|^p \, \mathrm{d}t \right) \mathrm{d}z$$
$$= C_p \int_{\mathbb{R}^n} |f(x)|^p \, \mathrm{d}x.$$

This, together with (5.8), proves the first assertion of the theorem. The other assertions easily follow by imitating the proof of the Lebesgue derivation theorem.

This computation is known under the name of "method of rotations."

5.3.2 ▪ General kernels

Theorem 5.17. *Let Ω be a 0-homogeneous function on $\mathbb{R}^n$ such that*

$$\int_{S_{n-1}} \Omega \, \mathrm{d}\sigma = 0 \qquad and \qquad \int_{S_{n-1}} |\Omega|^2 \mathrm{d}\sigma < \infty.$$

Let T be the convolution operator associated with the kernel $K(y) = \dfrac{\Omega(y)}{|y|^n}$. Then T is bounded on L^p for all $p \in (1, +\infty)$.

Proof. By decomposing Ω into its odd and even parts and using Theorem 5.16, we see that it is enough to consider the case of an even kernel. We consider the series expansion

$$\Omega = \sum_{k \geq 0} \alpha_k Y_k,$$

where Y_k is a normalized spherical harmonic of degree k and $\sum |\alpha_k|^2 < +\infty$. Then, due to Corollary 5.14, the Fourier transform of the kernel K is

$$\mathsf{m} = \sum_{k \geq 0} \gamma_{n,k} \alpha_k Y_k.$$

Now, consider the function $m_1(y) = \frac{y_1}{|y|} \mathsf{m}(y)$. It has the expansion $\sum_{k \geq 0} \gamma_{n,k} \alpha_k y_1 Y_k(y)$ (for $|y| = 1$). But we know (Lemma 5.10) that, for $k > 0$, $y_1 Y_k(y) = Y_{k-} + Y_{k+}$, where the terms on the right-hand side are spherical harmonics of degrees $k - 1$ and $k + 1$, respectively. Also, observe that since $\int_{S_{n-1}} |y_1 Y_k(y)|^2 \, \mathrm{d}\sigma(y) \leq 1$, we have

$$\|Y_{k-}\|_2 \leq 1 \qquad and \qquad \|Y_{k+}\|_2 \leq 1.$$

It follows that $y_1 \mathsf{m}(y) = \sum_{k > 0} Z_k(y)$, where Z_k is a spherical harmonic. There is no constant term since $y_1 m(y)$ is odd.

Since the ratios $\gamma_{n,k}/\gamma_{n,k+1}$ are bounded from above and from below, we have
$$\sum_{k>0} \gamma_{n,k}^{-2} \|Z_k\|_2^2 < +\infty.$$

If we set $\Omega_1^\flat = \sum_{k>0} \gamma_{n,k}^{-1} Z_k$, then the Fourier transform of the kernel $K_1^\flat(x) = \dfrac{\Omega^\flat(x/|x|)}{|x|^n}$ is $y_1 \mathsf{m}(y)$. It results from Theorem 5.16 that the corresponding operator $T_1^\flat$ is bounded on L^p for $1 < p < \infty$.

Of course, we have the same result for the operator $T_j^\flat$ whose multiplier is $y_j \mathsf{m}(y)$.

Now, consider the operator $U = \sum_{j=1}^{n} \mathrm{i} R_j T_j^\flat$, where R_j denotes the Riesz transforms. It is bounded on L^p for $1 < p < \infty$. However, by looking at Fourier transforms we see that, for $f \in L^2$, one has $Uf = Tf$. This means that $U = T$ and ends the proof.

5.4 ▪ Exercises

Exercise 5.1. *Determine the dimensions of $\mathscr{P}_{n,k}$ and $\mathscr{H}_{n,k}$.*

Exercise 5.2.

1. *Prove that $(x \pm \mathrm{i}y)^k$ are harmonic polynomials. Describe $\mathscr{H}_{2,k}$.*

2. *For $(x,y) \in \mathbb{R}^2 \setminus \{(0,0)\}$, one defines $\theta(x,y)$ by the equation $x + \mathrm{i}y = (x^2 + y^2)^{\frac{1}{2}} \mathrm{e}^{\mathrm{i}\theta(x,y)}$. Prove that the functions $\mathrm{e}^{\mathrm{i}k\theta(x,y)}$, for $k \in \mathbb{Z}$, generate the spherical harmonics on $\mathbb{R}^2$.*

3. *Let f be an integrable function on $\mathbb{R}^2$ such that*
$$f(x,y) = \varphi\left(\sqrt{x^2 + y^2}\right) \mathrm{e}^{\mathrm{i}n\theta(x,y)}.$$
Prove that the Fourier transform of f has the same form as f.

Exercise 5.3. *This is the generalization of part 3 of Exercise 5.2. Let $\mathcal{H}_\alpha$ stand for the Hilbert space of functions f on $\mathbb{R}^+$ such that $\|f\|_{\mathscr{H}_\alpha} = \left(\displaystyle\int_0^{+\infty} |f(r)|^2 r^{\alpha-1} \mathrm{d}r\right)^{1/2} < +\infty$. $(r = \sqrt{x^2 + y^2})$.*

1. *Let P be a solid spherical harmonic of degree k normalized so that $\displaystyle\int_{S_{n-1}} |P|^2 \mathrm{d}\sigma = 1$. Consider the Hilbert space*
$$\mathscr{R}_P = \left\{ x \mapsto f(|x|)P(x) \ : \ \int_{\mathbb{R}^n} \left| f(|x|)P(x) \right|^2 \mathrm{d}x < +\infty \right\}.$$
Prove that $\mathscr{R}_P$ and $\mathcal{H}_{2k+n-1}$ are isometrically isomorphic.

2. *Let $t > 0$. Compute the Fourier transform of $P(x)\mathrm{e}^{-\pi t|x|^2}$.*

3. *Show that for all $f \in \mathcal{H}_{n+2k}$, there exists $g \in \mathcal{H}_{n+2k}$ such that the following hold:*

   (a) *The Fourier transform of $f(|x|)P(x)$ is $(-\mathrm{i})^k g(|y|)P(y)$.*
       Hint: *Use Exercise* A.14.

   (b) *The correspondence $f \mapsto g$ is a unitary transformation of $\mathcal{H}_{n+2k}$.*

   These mappings are called Hankel transforms.

Chapter 6

BMO and H^1

Notation and conventions

If f is locally integrable on $\mathbb{R}^n$ and E is a measurable set of finite Lebesgue measure, we set

$$f_E = \frac{1}{|E|} \int_E f(x)\,\mathrm{d}x.$$

By cube in $\mathbb{R}^n$ we mean a cube whose sides are parallel to the coordinate axes. Such a cube will be designated by the symbol Q with or without indices.

6.1 ▪ The BMO space

Definitions 6.1.

1. *If f is locally integrable, we set*

$$f^\sharp(x) = \sup_{x \in Q} \frac{1}{|Q|} \int_Q |f(y) - f_Q|\,\mathrm{d}y$$

and

$$\|f\|_* = \sup_Q \frac{1}{|Q|} \int_Q |f(y) - f_Q|\,\mathrm{d}y = \|f^\sharp\|_\infty.$$

2. $\mathrm{BMO}(\mathbb{R}^n)$ *is the space of functions f such that* $\|f\|_* < \infty$.

The expression $\dfrac{1}{|Q|} \displaystyle\int_Q |f(y) - f_Q|\,\mathrm{d}y$ is the *mean oscillation* of f on Q, hence the name Bounded Mean Oscillation (BMO).

$\| \ \|_*$ is not a norm: $\|f\|_* = 0$ if and only if f is constant. Rather than a space of functions, BMO is a space of functions modulo constants. The verification that BMO is a Banach space is left to the reader.

$\| \ \|_*$ is invariant by translations and dilations: $x \mapsto f(x - a)$ and $x \mapsto f(\lambda x)$ have the same $\| \ \|_*$ norm as f.

For any $c \in \mathbb{C}$ and Q, we have $|f_Q - c| \leq \dfrac{1}{|Q|} \displaystyle\int_Q |f(y) - c|\, \mathrm{d}y$, so

$$\frac{1}{|Q|} \int_Q |f(y) - f_Q|\, \mathrm{d}y \leq \frac{2}{|Q|} \int_Q |f(y) - c|\, \mathrm{d}y.$$

Therefore, we have the following fact.

Proposition 6.2. $f \in \mathrm{BMO}$ *if and only if there exists a map* $Q \mapsto c_Q$ *such that*

$$\sup_Q \frac{1}{|Q|} \int_Q |f(y) - c_Q|\, \mathrm{d}y < +\infty.$$

It is clear that $L^\infty \subset \mathrm{BMO}$ and $\|\ \|_* \leq 2\|\ \|_\infty$. But these spaces are distinct; let us prove that the function $\log|\cdot|$ is in $\mathrm{BMO}(\mathbb{R})$.

As a function of a, $\displaystyle\int_{a-1/2}^{a+1/2} \big|\log|x|\big|\, \mathrm{d}x$ is continuous. Therefore

$$C = \sup_{|a| \leq 1} \int_{a-1/2}^{a+1/2} \big|\log|x|\big|\, \mathrm{d}x < +\infty,$$

and, for $h > 0$ and $|a| \leq h$,

$$\frac{1}{h} \int_{a-h/2}^{a+h/2} \big|\log|x| - \log h\big|\, \mathrm{d}x = \int_{a/h-1/2}^{a/h+1/2} \big|\log|x|\big|\, \mathrm{d}x \leq C.$$

For $|a| \geq 1$, we have $\displaystyle\int_{a-1/2}^{a+1/2} \big|\log|x| - \log|a|\big|\, \mathrm{d}x \leq \log\frac{|a| + 1/2}{|a| - 1/2}$. So,

$$\sup_{|a| \geq 1} \int_{a-1/2}^{a+1/2} \big|\log|x| - \log|a|\big|\, \mathrm{d}x \leq \log 3.$$

For $h > 0$ and $|a| \geq h$,

$$\frac{1}{h} \int_{a-h/2}^{a+h/2} \big|\log|x| - \log|a|\big|\, \mathrm{d}x = \int_{a/h-1/2}^{a/h+1/2} \big|\log|x| - \log|a/h|\big|\, \mathrm{d}x \leq \log 3.$$

It is then easy to prove, using Proposition 6.2, that $\log|\cdot|$ is in $\mathrm{BMO}(\mathbb{R})$.

We can define on BMO another norm equivalent to $\|\ \|_*$:

$$\|f\|_{\mathrm{BMO}} = \sup_Q \inf_{c \in \mathbb{C}} \frac{1}{|Q|} \int_Q |f(y) - c|\, \mathrm{d}y.$$

Definition 6.3 (dyadic BMO). *Let* $\mathcal{D}$ *be a family of dyadic cubes. When using only dyadic cubes in* $\mathscr{D}$ *instead of any cubes, one defines* $f_\mathcal{D}^\sharp$ *instead of* $f^\sharp$. *The space* $\mathrm{BMO}_\mathcal{D}$ *is the space of functions* f *such that* $f_\mathcal{D}^\sharp \in L^\infty$. *Of course, the sharp function depends on the choice of* $\mathcal{D}$.

Obviously, $\mathrm{BMO} \subset \mathrm{BMO}_\mathcal{D}$. The reader is invited to check that the function $\log|\cdot|\,\mathbf{1}_{(0,+\infty)}$ is in $\mathrm{BMO}_\mathcal{D}(\mathbb{R})$, when using standard dyadic cubes (see Section 3.1), but not in $\mathrm{BMO}(\mathbb{R})$.

Lemma 6.4 (good lambda inequality). *Let f and g be two positive and measurable functions on a measure space $(X, \mathcal{A}, \mu)$ and let p be a positive number. Assume that there exist three constants a, b, and c such that $ab^{-p} < 1$ and such that, for all $\lambda > 0$, one has*

$$\mu(\{f > \lambda, \, g \leq c\lambda\}) \leq a\,\mu(\{f > b\lambda\}).$$

Then

1. *if $\|f\|_p < +\infty$, then $\|f\|_p^p \leq \dfrac{c^{-p}}{1 - ab^{-p}}\,\|g\|_p^p$;*

2. *if $b < 1$ and if there exists $p_0 \in (0, p)$ such that $\|f\|_{p_0} < +\infty$, then $\|f\|_p^p \leq \dfrac{c^{-p}}{1 - ab^{-p}}\,\|g\|_p^p$.*

Proof. We have $\mu(\{f > \lambda\}) \leq a\,\mu(\{f > b\lambda\}) + \mu(\{g > c\lambda\})$.

In the first case,

$$
\begin{aligned}
\|f\|_p^p &= \int_0^{+\infty} p\lambda^{p-1}\mu(\{f > \lambda\})\,\mathrm{d}\lambda \\
&\leq b^{-p}\int_0^{+\infty} ap\lambda^{p-1}\mu(\{f > \lambda\})\,\mathrm{d}\lambda + c^{-p}\int_0^{+\infty} p\lambda^{p-1}\mu(\{g > \lambda\})\,\mathrm{d}\lambda \\
&= ab^{-p}\|f\|_p^p + c^{-p}\|g\|_p^p,
\end{aligned}
$$

which gives the result since we assumed that $\|f\|_p$ is finite and $ab^{-p} < 1$.

In the second case, for $R > 0$ we have

$$
\begin{aligned}
\int_0^R & p\lambda^{p-1}\mu(\{f > \lambda\})\,\mathrm{d}\lambda \\
&\leq ab^{-p}\int_0^{bR} p\lambda^{p-1}\mu(\{f > \lambda\})\,\mathrm{d}\lambda + c^{-p}\int_0^{cR} p\lambda^{p-1}\mu(\{g > \lambda\})\,\mathrm{d}\lambda \\
&\leq ab^{-p}\int_0^{R} p\lambda^{p-1}\mu(\{f > \lambda\})\,\mathrm{d}\lambda + c^{-p}\int_0^{+\infty} p\lambda^{p-1}\mu(\{g > \lambda\})\,\mathrm{d}\lambda,
\end{aligned}
$$

which yields the result since $\int_0^R p\lambda^{p-1}\mu(\{f > \lambda\})\,\mathrm{d}\lambda$ is finite (do not forget that $\|f\|_{p_0}$ is finite). $\blacksquare$

Lemma 6.5. *Let f be a measurable function on $\mathbb{R}^n$. For any $b \in (0, 1)$ and $c > 0$, we have*

$$\left|\{f_{\mathcal{D}}^{\sharp} \leq c\lambda, \, \mathsf{M}_{\mathcal{D}}f > \lambda\}\right| \leq \frac{2^n c}{1 - b}\left|\{\mathsf{M}_{\mathcal{D}}f > b\lambda\}\right|$$

for all $\lambda > 0$ ($\mathsf{M}_{\mathcal{D}}$ stands for the dyadic maximal operator).

Proof. Fix $b < 1$. If $|\{\mathsf{M}_{\mathcal{D}}f > b\lambda\}| = +\infty$, there is nothing to prove. Otherwise, we may write

$$\{\mathsf{M}_{\mathcal{D}}f > b\lambda\} = \bigcup \mathsf{Q}_j,$$

where the Q_j are dyadic cubes whose interior are disjoint and such that $b\lambda < |f|_{\mathsf{Q}_j} \leq 2^n b\lambda$ (these cubes are those which are maximal for inclusion among the Q such that $|f|_{\mathsf{Q}} > b\lambda$).

Let Q be one of the Q_j and let $\widetilde{Q}$ be its predecessor. We have $|f|_{\widetilde{Q}} \leq b\lambda$, due to the maximality of Q.

Consider $x \in Q$ such that $\mathsf{M}_{\mathcal{D}}f(x) > \lambda$. Since for dyadic cubes Q^* containing $\widetilde{Q}$ we have $|f|_{Q^*} \leq b\lambda < \lambda$, so

$$\mathsf{M}_{\mathcal{D}}f(x) = \sup_{x \in Q^* \subset Q} \frac{1}{|Q^*|} \int_{Q^*} |f(y)| \, \mathrm{d}y.$$

However, if $x \in Q^* \subset Q$, we have

$$\frac{1}{|Q^*|} \int_{Q^*} |f(y) - f_{\widetilde{Q}}| \mathbf{1}_Q(y) \, \mathrm{d}y \geq \frac{1}{|Q^*|} \int_{Q^*} |f(y)| \, \mathrm{d}y - |f_{\widetilde{Q}}|$$

$$\geq \frac{1}{|Q^*|} \int_{Q^*} |f(y)| \, \mathrm{d}y - b\lambda.$$

This proves that $\mathsf{M}_{\mathcal{D}}\big((f - f_{\widetilde{Q}}) \mathbf{1}_Q\big)(x) \geq (1-b)\,\lambda$ and

$$Q \cap \{\mathsf{M}_{\mathcal{D}}f > \lambda\} \subset \Big\{\mathsf{M}_{\mathcal{D}}\big((f - f_{\widetilde{Q}}) \mathbf{1}_Q\big) \geq (1-b)\,\lambda\Big\}.$$

So, we have

$$\big|Q \cap \{\mathsf{M}_{\mathcal{D}}f > \lambda\}\big| \leq \frac{1}{(1-b)\,\lambda} \int_Q |f(y) - f_{\widetilde{Q}}| \, \mathrm{d}y \leq \frac{2^n |Q|}{(1-b)\,\lambda} \, f_{\mathcal{D}}^{\sharp}(z)$$

for any $z \in Q$. As a consequence, if there exists $z \in Q$ such that $f_{\mathcal{D}}^{\sharp}(z) \leq c\lambda$, we have

$$\big|Q \cap \{\mathsf{M}_{\mathcal{D}}f > \lambda\}\big| \leq \frac{2^n c \, |Q|}{1 - b}.$$

It results that

$$\big|Q \cap \{f^{\sharp} \leq c\lambda, \mathsf{M}_{\mathcal{D}}f > \lambda\}\big| \leq \frac{2^n c \, |Q|}{1 - b}$$

and

$$\big|\{f^{\sharp} \leq c\lambda, \mathsf{M}_{\mathcal{D}}f > \lambda\}\big| \leq \sum_j \big|Q_j \cap \{f^{\sharp} \leq c\lambda, \mathsf{M}_{\mathcal{D}}f > \lambda\}\big|$$

$$\leq \sum_j \frac{2^n c \, |Q_j|}{1 - b} = \frac{2^n c}{1 - b} \, \big|\{\mathsf{M}_{\mathcal{D}}f > b\lambda\}\big|.$$

Corollary 6.6. *If $1 < p < +\infty$, there exists a constant $C_{p,n}$ such, for all $f \in L^p(\mathbb{R}^n)$, one has $\|\mathsf{M}_{\mathcal{D}}f\|_p \leq C_{p,n}\|f_{\mathcal{D}}^{\sharp}\|_p$.*

Theorem 6.7 (John–Nirenberg inequality). *There exist c_1 and c_2 such that, for every cube Q, for all $\lambda > 0$, and for all f such that $\|f\|_* \neq 0$, one has*

$$|Q \cap \{|f - f_Q| > \lambda\}| \leq c_1 e^{-c_2 \lambda / \|f\|_*} |Q|.$$

Proof. Let Q be an arbitrary cube. Consider the family of dyadic cubes subordinate to Q (see Section 3.1). Set $\varphi = (f - f_Q) \mathbf{1}_Q$. If Q^* is a cube containing Q, one has $\varphi_{Q^*} = 0$, so

$$\frac{1}{|Q^*|} \int_{Q^*} |\varphi(y) - \varphi_{Q^*}| \, \mathrm{d}y = \frac{|Q|}{|Q^*|} \frac{1}{|Q|} \int_Q |f(y) - f_Q| \, \mathrm{d}y \leq \|f\|_*. \tag{6.1}$$

It results that if $\varphi^\sharp_\mathcal{D}$ stands for the $\sharp$ function associated with the dyadic cubes based on Q, we have $\varphi^\sharp_\mathcal{D} \le \|f\|_*$. Now, for any $t > 0$, set $\lambda = t + 2^{n+1}\|f\|_*$, take $c = \lambda^{-1}\|f\|_*$, b such that $\dfrac{2^n c}{1-b} = \dfrac{1}{2}$, and use Lemma 6.5. We obtain

$$|\{\mathsf{M}_\mathcal{D}\varphi > \lambda\}| \le \frac{1}{2}\,|\{\mathsf{M}_\mathcal{D}\varphi > \lambda - 2^{n+1}\|f\|_*\}|,$$

i.e.,

$$|\{\mathsf{M}_\mathcal{D}\varphi > t + 2^{n+1}\|f\|_*\}| \le \frac{1}{2}\,|\{\mathsf{M}_\mathcal{D}\varphi > t\}|.$$

Also, due to (6.1), we have $|\{\mathsf{M}_\mathcal{D}\varphi > \|f\|_*\}| \le |Q|$. If we set $\gamma(t) = |\{\mathsf{M}_\mathcal{D}\varphi > t\}|$ for all $t > 0$, we have $\gamma(2^{n+1}\|f\|_*) \le |Q|$ and $\gamma(t + 2^{n+1}\|f\|_*) \le \frac{1}{2}\gamma(t)$, which implies the existence of positive numbers c_1 and c_2 such that

$$\gamma(t) \le c_1 \mathrm{e}^{-c_2 t/\|f\|_*}|Q|.$$

To end the proof, just notice that $|\varphi| \le \mathsf{M}_\mathcal{D}\varphi$ almost everywhere.

Corollary 6.8. *For any $p \in (1, +\infty)$,*

$$f \longmapsto \sup_Q \left(\frac{1}{|Q|} \int_Q |f(y) - f_Q|^p \,\mathrm{d}y \right)^{1/p}$$

defines on BMO *a norm equivalent to* $\|\cdot\|_*$.

Proof. On the one hand, the Hölder inequality gives

$$\frac{1}{|Q|} \int_Q |f(y) - f_Q|\,\mathrm{d}y \le \left(\frac{1}{|Q|} \int_Q |f(y) - f_Q|^p \,\mathrm{d}y \right)^{1/p}.$$

On the other hand, the John–Nirenberg inequality gives

$$\frac{1}{|Q|} \int_Q |f(y) - f_Q|^p \,\mathrm{d}y \le c_1 p \int_0^{+\infty} \lambda^{p-1} \mathrm{e}^{-c_2\lambda/\|f\|_*}\,\mathrm{d}\lambda$$
$$= C\,\|f\|_*^p.$$

6.2 ▪ The $H^1(\mathbb{R}^n)$ space

If E is a subset of $\mathbb{R}^n$ of finite Lebesgue measure and f a measurable function, we set

$$\|f\|_{L^p(E)} = \left(\frac{1}{|E|} \int_E |f(x)|^p \mathrm{d}x \right)^{1/p} \quad \text{and} \quad \|f\|_{L^\infty(E)} = \operatorname{ess\,sup} f \mathbf{1}_E.$$

Definition 6.9. *Let $p \in (1, +\infty]$. A p-atom is a function a such that*

1. *a vanishes outside a cube Q;*

2. *$\displaystyle\int a(x)\,\mathrm{d}x = 0$;*

3. *$\|a\|_{L^p(Q)} \le \dfrac{1}{|Q|}$ (i.e., $\left(\displaystyle\int |a(x)|^p \mathrm{d}x \right)^{1/p} \le |Q|^{-1/p'}$ when $p < \infty$).*

Definition 6.10. *For $p \in (1, +\infty]$,*

1. $H^{1,p}(\mathbb{R}^n) = \left\{ \sum \lambda_j a_j \; : \; \sum |\lambda_j| < +\infty, \quad a_j \text{ being a } p\text{-atom} \right\}$;

2. *for $f \in H^{1,p}(\mathbb{R}^n)$*

$$\|f\|_{1,p} = \inf \left\{ \sum |\lambda_j| \; : \; f = \sum \lambda_j a_j, \quad a_j \text{ being a } p\text{-atom} \right\}.$$

These spaces are Banach spaces. The Hölder inequality implies that for any p-atom a, $\|a\|_1 \leq 1$, and that if a is a q-atom, it is a p-atom for all $p \in (1, q)$. So we have the following inequalities:

1. $\|f\|_{L^1(\mathbb{R}^n)} \leq \|f\|_{1,p}$ for $1 < p \leq +\infty$.

2. $\|f\|_{1,p} \leq \|f\|_{1,q}$ for $1 < p \leq q \leq +\infty$.

The notation $H^{1,p}$ is provisional because, as we shall see, all these spaces are isomorphic. After having proven this, all these spaces will be denoted by $H^1(\mathbb{R}^n)$.

Lemma 6.11. *For any $p \in (1, +\infty)$ there exists K_p such that, for any p-atom a, we have $a = g + b$, with*

$$\|g\|_{1,\infty} \leq K_p \quad and \quad \|b\|_{1,p} \leq \frac{1}{4}.$$

Proof. Let a be a p-atom associated with a cube Q. Let t be a number larger than $1/|Q|$ which will be chosen later on. Let us consider a Calderón–Zygmund decomposition: the dyadic cubes we consider are those based on Q (see Section 3.1). The cubes Q_j are the dyadic subcubes of Q which are maximal among those which satisfy $\left(\frac{1}{|Q'|} \int_{Q'} |a(x)|^p \, dx \right)^{\frac{1}{p}} > t$. We have $\|a\|_{L^p(Q_j)} \leq 2^{n/p} t$. One sets $\Omega = \bigcup Q_j$,

$$g(x) = \begin{cases} a(x) & \text{if } x \notin \Omega \\ a_{Q_j} & \text{if } x \in Q_j \end{cases}, \quad b_j = (a - a_{Q_j}) \mathbf{1}_{Q_j}, \quad \text{and} \quad b = \sum b_j.$$

The function g is essentially bounded by $2^{n/p} t$, vanishes outside Q, and its integral is zero. This means that $2^{-n/p} t^{-1} |Q|^{-1} g$ is an ∞-atom. Therefore, $\|g\|_{1,\infty} \leq 2^{n/p} t |Q|$. We have

$$\|b_j\|_{L^p(Q_j)} \leq \|a\|_{L^p(Q_j)} + |a_{Q_j}| \leq 2 \|a\|_{L^p(Q_j)}.$$

As b_j is of mean 0 and is supported in Q_j, it is a multiple of a p-atom and we have

$$\|b_j\|_{1,p} \leq 2 |Q_j| \|a\|_{L^p(Q_j)} = 2 |Q_j|^{\frac{1}{p'}} \left(\int_{Q_j} |a(x)|^p \, dx \right)^{1/p}$$

and, by using Hölder inequality,

$$\|b\|_{1,p} \le 2 \left(\sum |Q_j| \right)^{1/p'} \left(\sum \int_{Q_j} |a(x)|^p \mathrm{d}x \right)^{1/p}$$

$$\le 2 \left(t^{-p} \sum \int_{Q_j} |a(x)|^p \, \mathrm{d}x \right)^{1/p'} \left(\int_\Omega |a(x)|^p \mathrm{d}x \right)^{1/p}$$

$$\le 2t^{-p/p'} \|a\|_p^p \le \frac{2}{(t\,|Q|)^{p/p'}}.$$

We get the result by choosing $t = 2^{3p'/p}|Q|^{-1}$.

Corollary 6.12. *Fix $p \in (1, +\infty)$. Any $f \in H^{1,p}(\mathbb{R}^n)$ can be written as $f = g + h$, where $\|g\|_{1,\infty} \le 2K_p\|f\|_{1,p}$ and $\|h\|_{1,p} \le \frac{1}{2} \|f\|_{1,p}$.*

Proof. Write $f = \sum \lambda_j a_j$, with $\sum |\lambda_j| \le 2\|f\|_{1,p}$, apply Lemma 6.11 to each p-atom a_j, and regroup the corresponding terms.

Theorem 6.13. *For all $p \in (1, +\infty)$, $H^{1,p}$ is isomorphic to $H^{1,\infty}$.*

Proof. We are given $f \in H^{1,p}$. Apply Corollary 6.12: $f = g_1 + f_1$, with $\|g_1\|_{1,\infty} \le 2K_p\|f\|_{1,p}$ and $\|f_1\|_{1,p} \le \frac{1}{2} \|f\|_{1,p}$. Still using Corollary 6.12, we define by recursion two sequences: $f_j = g_{j+1} + f_{j+1}$, with $\|g_{j+1}\|_{1,\infty} \le 2K_p\|f_j\|_{1,p}$ and $\|f_{j+1}\|_{1,p} \le \frac{1}{2} \|f_j\|_{1,p}$. Then we have

$$\|f_j\|_{1,p} \le 2^{-j}\|f\|_{1,p}, \quad \|g_j\|_{1,\infty} \le 2^{-j}K_p\|f\|_{1,p}, \quad \text{and } f = g_1 + \cdots + g_k + f_k.$$

This proves that $g_1 + \cdots + g_k$ has a limit in $H^{1,\infty}$ and converges towards f in $H^{1,p}$. Since both these convergences imply convergence in L^1, f is also the limit of $g_1 + \cdots + g_k$ in $H^{1,\infty}$. One concludes by using the Banach isomorphism theorem.

6.3 ▪ Duality of H^1–BMO

Theorem 6.14. *The space $\mathrm{BMO}(\mathbb{R}^n)$ is the dual space of $H^1(\mathbb{R}^n)$.*

Proof. Let us first consider $g \in \mathrm{BMO}$. To show that it defines a continuous linear form on H^1 we use the definition of H^1 by mean of p-atoms. So, let a be a p-atom associated with a cube Q, with $1 < p < +\infty$. We have

$$\int_{\mathbb{R}^n} g(x)a(x) \, \mathrm{d}x = \int_Q g(x)a(x) \, \mathrm{d}x = \int_Q (g(x) - g_Q) \, a(x) \, \mathrm{d}x.$$

So, by using the Hölder and John–Nirenberg inequalities, we get

$$\left| \int_{\mathbb{R}^n} g(x)a(x) \, \mathrm{d}x \right| \le \|a\|_p \left(\int_Q |g(x) - g_Q|^{p'} \mathrm{d}x \right)^{1/p'}$$

$$\le \left(\frac{1}{|Q|} \int_Q |g(x) - g_Q|^{p'} \right)^{1/p'} \le C_p \|g\|_*.$$

As the p-atoms generate H^1, this proves that any element of BMO defines a continuous linear form on H^1.

Conversely, let T be a bounded linear form on H^1. Let $T_\mathbb{R}$ stand for the restriction of T to $H^1_\mathbb{R}$, the space of real elements in H^1; this is an $\mathbb{R}$-linear map from $H^1_\mathbb{R}$ to $\mathbb{C}$. If $f = g + \mathrm{i}\,h \in H^1$, with g and h in $H^1_\mathbb{R}$, we have

$$T(f) = T(g) + \mathrm{i}\,T(h) = T_\mathbb{R}(g) + \mathrm{i}\,T_\mathbb{R}(h).$$

We can further decompose $T_\mathbb{R}$: Define, for $f \in H^1_\mathbb{R}$, $T_1 f = \Re T_\mathbb{R}(f)$ and $T_2 = \Im T_\mathbb{R}(f)$. Then T_1 and T_2 are elements of the dual space to the real Banach space $H^1_\mathbb{R}$. If we show that $T_j(f) = \int f(x)\varphi_j(x)\,\mathrm{d}x$, for suitable elements φ_1 and φ_2 of BMO, we will have

$$T(g + \mathrm{i}\,h) = \int \big(\varphi_1(x) + \mathrm{i}\,\varphi_2(x)\big)\,g(x)\,\mathrm{d}x + \mathrm{i}\int \big(\varphi_1(x) + \mathrm{i}\,\varphi_2(x)\big)\,h(x)\,\mathrm{d}x$$

$$= \int \big(\varphi_1(x) + \mathrm{i}\,\varphi_2(x)\big)\big(g(x) + \mathrm{i}\,h(x)\big)\,\mathrm{d}x.$$

So we may consider an element T of the dual space to $H^1_\mathbb{R}$. Let us set $\mathsf{Q}_k = [-k, k]^n$ and consider the real Hilbert space H_k of real square integrable functions supported in Q_k and of mean value 0. Each element f of H_k is proportional to a 2-atom; it follows that $\|f\|_{1,2} \le \|f\|_{L^2(\mathsf{Q}_k)}|\mathsf{Q}_k|$. It results that T defines a continuous linear form on H_k, so there exists $\psi_k \in L^2(\mathsf{Q}_k)$, unique up to an additive constant, such that, for any $f \in \mathsf{H}_k$, we have $T(f) = \int f(x)\psi_k(x)\,\mathrm{d}x$. This last equality holds in particular when $f \in L^2(\mathsf{Q}_1)$. This means that we can modify ψ_k by adding to it the indicator function of Q_k multiplied by a suitable constant so that the restriction of ψ_k to Q_1 be ψ_1. Then the restriction of ψ_{k+1} to Q_k coincides with ψ_k for all k. Therefore there is a function φ on $\mathbb{R}^n$ whose restriction to each Q_k is ψ_k. It results that, for any atom a, one has $T(a) = \int a(x)\,\varphi(x)\,\mathrm{d}x$.

Now, we wish to show that φ is in BMO. Let Q be an arbitrary cube. Let c_Q be a median of φ on Q. This means that c_Q is a number such that $|\{\varphi < c_\mathsf{Q}\} \cap \mathsf{Q}| \le |\mathsf{Q}|/2$ and $|\{\varphi > c_\mathsf{Q}\} \cap \mathsf{Q}| \le |\mathsf{Q}|/2$. Since the Lebesgue measure has no atom, we can find (see Exercise 6.7) a measurable subset E of Q such that

$$|E| = |\mathsf{Q}|/2, \quad \{\varphi > c_\mathsf{Q}\} \cap \mathsf{Q} \subset E, \quad \text{and} \quad \{\varphi < c_\mathsf{Q}\} \cap \mathsf{Q} \subset \mathsf{Q} \setminus E.$$

The function $a = \dfrac{1}{|\mathsf{Q}|}\left(\mathbf{1}_{\mathsf{Q} \setminus E} - \mathbf{1}_E\right)$ is an ∞-atom. We have

$$T(a) = \int_\mathsf{Q} a(x)\varphi(x)\,\mathrm{d}x = \int_\mathsf{Q} a(x)\big(\varphi(x) - c_\mathsf{Q}\big)\,\mathrm{d}x$$

$$= \frac{1}{|\mathsf{Q}|}\int_\mathsf{Q} |\varphi(x) - c_\mathsf{Q}|\,\mathrm{d}x,$$

so

$$\frac{1}{|\mathsf{Q}|}\int_\mathsf{Q} |\varphi(x) - c_\mathsf{Q}|\,\mathrm{d}x \le \|T\|,$$

where $\|T\|$ is the norm of T when H^1 is endowed with the norm $\|\cdot\|_{1,\infty}$. It results that φ is in BMO. This ends the proof.

Theorem 6.15. *Let K be a square integrable function on $\mathbb{R}^n$ such that*

1. $\|\widehat{K}\|_\infty \le B < +\infty$;

2. $\displaystyle\sup_{y \ne 0} \int_{|x| \ge 2|y|} |K(x-y) - K(x)| \, \mathrm{d}x \le B.$

Then the operator T (defined by $Tf = K * f$) is bounded from $H^1(\mathbb{R}^n)$ to $L^1(\mathbb{R}^n)$.

Proof. Due to the invariance by translations it is enough to prove that there exists M such that, for any 2-atom a on a cube Q centered at the origin, we have $\|K * a\|_{L^1} \le M$.

Let Q* be Q dilated by the ratio $2\sqrt{n}$. We have

$$\int_{Q^*} |Ta(x)| \, \mathrm{d}x \le |Q^*|^{1/2} \|Ta\|_2 \le (2\sqrt{n})^{n/2} B \, |Q|^{1/2} \|a\|_2 \le (2\sqrt{n})^{n/2} B.$$

Now we have to estimate $\displaystyle\int_{^cQ^*} |Ta(x)| \, \mathrm{d}x$. We have

$$\int_{^cQ^*} |Ta(x)| \, \mathrm{d}x = \int_{^cQ^*} \left| \int_Q \left(K(x-y) - K(x) \right) a(y) \, \mathrm{d}y \right| \mathrm{d}x$$

$$\le \int_Q |a(y)| \left(\int_{^cQ^*} |K(x-y) - K(x)| \, \mathrm{d}x \right) \mathrm{d}y$$

$$\le B \int |a(y)| \, \mathrm{d}y \le B.$$

This provides another proof of the boundedness on L^p of Calderón–Zygmund operators. Indeed, since such an operator is bounded from H^1 to L^1, by duality it is bounded from L^∞ to BMO. Then Theorem 8.5 shows that it is bounded on L^p for $2 \le p < +\infty$ and, again by duality, for $1 < p \le 2$.

6.4 ▪ Exercises

Exercise 6.1. *All intervals are semiopen to the right.*

1. *Prove that there exists a constant γ such that, for all nondyadic intervals I, there are two contiguous dyadic intervals J_1 and J_2, such that $|J_1| = |J_2|$, $I \subset J_1 \cup J_2$ and $\frac{1}{\gamma} |J_1| \le |I| \le \gamma |J_1|$.*

2. *Let f be a function in $\mathrm{BMO}_\mathcal{D}(\mathbb{R})$. Suppose that there is a constant $C > 0$ such that, for any pair (J_1, J_2) of adjacent dyadic intervals of the same length, one has $|f_{J_1} - f_{J_2}| \le C$.*
Show that $f \in \mathrm{BMO}(\mathbb{R})$.

Exercise 6.2. *Let φ be the function so defined on $\mathbb{R}$:*

$$\varphi(x) = \begin{cases} 0 & \text{if } x \le 0, \\ \log|x| & \text{if } x > 0. \end{cases}$$

Show that φ is in $\mathrm{BMO}_\mathcal{D}(\mathbb{R})$ but not in $\mathrm{BMO}(\mathbb{R})$.

Exercise 6.3. *Find two functions in $\mathrm{BMO}(\mathbb{R})$ whose product is not a BMO function.*

Exercise 6.4.

1. *Let f be an element of* $\mathrm{BMO}(\mathbb{R}^n)$, *and let Q be a cube centered at zero.*

   (a) *Let λ be a real number larger than 1. Show the following inequalities:*
   $$|f_Q - f_{\lambda Q}| \le \lambda^n \|f\|_{\mathrm{BMO}},$$
   $$|f_Q - f_{\lambda^k Q}| \le k\lambda^n \|f\|_{\mathrm{BMO}} \quad \text{for any integer } k \ge 0.$$

   Hint: *Write $f_Q - f_{\lambda Q} = |Q|^{-1} \int_Q (f(t) - f_{\lambda Q}) \, dt$.*

   (b) *Let a be a real number larger than 1. Show the following inequalities*
   $$|f_Q - f_{aQ}| \le ka^{n/k} \|f\|_{\mathrm{BMO}} \quad \text{for any integer } k \ge 1,$$
   $$|f_Q - f_{aQ}| \le e(n \log a + 1) \|f\|_{\mathrm{BMO}}.$$

   Hint: *To get the second inequality from the first one, note that if k were allowed to be any real number, the minimum of $ka^{n/k}$ would be attained for $k = n \log a$.*

   (c) *Let a be a positive number. Show that*
   $$|f_Q - f_{aQ}| \le e(n|\log a| + 1) \|f\|_{\mathrm{BMO}}.$$

2. *Let δ be a positive number. Does the function $x \mapsto \left| \log |x| \right|^{1+\delta}$ belong to $\mathrm{BMO}(\mathbb{R})$?*

Exercise 6.5. *Let f be a locally integrable function on $\mathbb{R}^2$ assuming real values only. One assumes that there exists a constant C_0 such that the following hold:*

- *for all x, the function $y \mapsto f(x,y)$ lies in BMO with a norm less than or equal to C_0;*

- *for all y, the function $x \mapsto f(x,y)$ lies in BMO with a norm less than or equal to C_0.*

1. *Show that if $Q = I \times J$ is a square in $\mathbb{R}^2$, one has*
   $$\frac{1}{|J|} \int_J \left| \frac{1}{|I|} \int_I f(x,y) \, dx - \frac{1}{|Q|} \int_Q f(x,t) \, dx \, dt \right| dy \le C_0.$$

2. *Deduce from this that $\|f\|_{\mathrm{BMO}(\mathbb{R}^2)} \le 2 C_0$.*

3. *Let P be a polynomial in two variables with complex coefficients. Show that $\log |P|$ is in BMO and give an upper bound of its norm.*

Exercise 6.6. *For $\alpha > 0$, set $g_\alpha(x) = \log(x^2 + \alpha)$.*

1. *Show that, for all a and b such that $0 \le a \le b$, one has*
   $$\int_a^b \left(\log(1 + b^2) - \log(1 + x^2) \right) dx \le 2(b - a).$$

   Conclude that $g_1 \in \mathrm{BMO}$.

2. *Show that, for all $\alpha > 0$, one has $\|g_\alpha\|_{\mathrm{BMO}} = \|g_1\|_{\mathrm{BMO}}$. Deduce from this fact that $\|g_\alpha\|_{\mathrm{BMO}} = 2\,\|\log|x|\|_{\mathrm{BMO}}$.*

3. *Set $c = \|\log|x|\|_{\mathrm{BMO}}$. Show that if P is a polynomial of degree n with complex coefficients, then $\log|P|$ is in BMO with a norm not exceeding cn (factor P and use the preceding question).*

Exercise 6.7. *Let $(X, \mathscr{A}, \mu)$ be a probability space with no atom. If f is a nonnegative measurable function, its distribution function is so defined:*

$$\varrho_f(t) = \mu(f > t) \qquad \big(t \in (0, +\infty)\big).$$

The function ϱ_f is nonincreasing, continuous to the right, and has limits to the left $\big(\varrho_f(t_-) = \mu(f \geq t)\big)$.
   Show that there exist $E \in \mathscr{A}$ and $c > 0$ such that

$$\mu(E) = \frac{1}{2}, \qquad \{f > c\} \subset E, \qquad \text{and} \qquad \{f < c\} \subset {}^cE.$$

Hint: *Consider two cases.*

1. *There exists c such that $\varrho_f(c) = \varrho_f(c_-) = 1/2$.*

2. *There exists c such that $\varrho_f(c) \leq 1/2 \leq \varrho_f(c_-)$ and $m_f(c) < m_f(c_-)$.*

Exercise 6.8. *Prove that $\mathrm{BMO}(\mathbb{R}^n)$ is complete.*

Exercise 6.9.

1. *Let $(X, \mathscr{A}, \mu)$ be a measure space. If f is a measurable function from X to $\mathbb{C}$, define its distribution function ϱ_f,*

$$\varrho_f(t) = \mu(\{|f| > t\}) \qquad for \quad t > 0,$$

and its nonincreasing rearrangement f^*,

$$f^*(t) = \inf\{u \,:\, \varrho_f(u) \leq t\} \qquad for \quad t \in (0, +\infty).$$

Show that, for all $t > 0$, $|\{f^* > t\}| = \varrho_f(t)$.

2. *Now suppose that μ is a probability measure. Let f be a real and μ-integrable function. Show that*

$$\int_X |f(x) - c|\,\mathrm{d}x$$

is minimum if c is a median for f relatively to μ.

Hint: *Consider first the case when f is a nonincreasing real function on the interval $[0, 1]$ endowed with the Lebesgue measure.*

Exercise 6.10. *Prove that the spaces $H^{1,p}$ are complete.*

Exercise 6.11. *Show that $(\log|x|)^2$ is not in $\mathrm{BMO}(\mathbb{R})$.*
Hint: *This function does not fulfill the John–Nirenberg inequality.*

Chapter 7

Singular integrals on other groups

7.1 ▪ The torus

In this section we transfer results of harmonic analysis on the real line to the torus $\mathbb{R}/2\pi\mathbb{Z}$. Exercise 7.1 as well as the next section give other examples of transfers from one group to another.

7.1.1 ▪ Background

Let $\mathbb{T}$ be the torus $\mathbb{R}/2\pi\mathbb{Z}$. A function on $\mathbb{T}$ is identified with a 2π-periodic function on $\mathbb{R}$. The characters of the group $\mathbb{T}$ are the functions e^{ijx} with $j \in \mathbb{Z}$. For $1 \le p < +\infty$, the space $L^p(\mathbb{T})$ is the set of (classes modulo almost everywhere equality of) measurable functions f such that their norm

$$\|f\|_{L^p(\mathbb{T})} = \left(\int_0^{2\pi} |f(t)|^p \, \frac{dt}{2\pi} \right)^{\frac{1}{p}}$$

is finite.

The Fourier transform of $f \in L^1$ is the sequence $\hat{f} = \left(\hat{f}(j) \right)_{j \in \mathbb{Z}}$, where

$$\hat{f}(j) = \int_0^{2\pi} f(t) \, e^{-ijt} \, \frac{dt}{2\pi}.$$

A trigonometric polynomial is a finite sum of the form $P(x) = \sum c_j e^{ijx}$. Then $c_j = \hat{P}(j)$.

If f is integrable and P is a trigonometric polynomial, then $f * g(x) = \sum \hat{P}(j) \, \hat{f}(j) \, e^{ijx}$.

7.1.2 ▪ Conjugate function

We define the sgn function:

$$\mathrm{sgn}(j) = \begin{cases} +1 & \text{if } j > 0, \\ 0 & \text{if } j = 0, \\ -1 & \text{if } j < 0. \end{cases}$$

75

Let us consider the following trigonometric polynomials:

$$D_m^*(x) = \sum_{1 \leq |j| \leq n} -\frac{\mathrm{i}\,\mathrm{sgn}(j)}{2}\,\mathrm{e}^{\mathrm{i}jx} = \frac{1}{2 \tan \frac{x}{2}} - \frac{\cos\left(m + \frac{1}{2}\right)x}{2 \sin \frac{x}{2}}. \tag{7.1}$$

Let φ be a C^1 2π-periodic function. We have

$$\int_{-\pi}^{\pi} \varphi(x)\,D_m^*(x)\,\mathrm{d}x = \int_{-\pi}^{\pi} \left(\varphi(x) - \varphi(0)\right) D_m^*(x)\,\mathrm{d}x$$

$$= \int_{-\pi}^{\pi} \frac{\varphi(x) - \varphi(0)}{2 \tan \frac{x}{2}}\,\mathrm{d}x - \int_{-\pi}^{\pi} \frac{\varphi(x) - \varphi(0)}{2 \sin \frac{x}{2}} \cos\left(\left(m + \frac{1}{2}\right)x\right)\,\mathrm{d}x.$$

The function $\left(\varphi(x) - \varphi(0)\right)/\sin \frac{x}{2}$ being integrable, the Riemann–Lebesgue theorem yields

$$\lim_{m \to +\infty} \int_{-\pi}^{\pi} \varphi(x)\,D_m^*(x)\,\mathrm{d}x = \int_{-\pi}^{\pi} \frac{\varphi(x) - \varphi(0)}{2 \tan \frac{x}{2}}\,\mathrm{d}x.$$

Since $\left(\varphi(x) - \varphi(0)\right)/\tan \frac{x}{2}$ is integrable, we also have

$$\lim_{m \to +\infty} \int_{-\pi}^{\pi} \varphi(x)\,D_m^*(x)\,\mathrm{d}x = \lim_{\varepsilon \to 0} \int_{\varepsilon \leq |x| \leq \pi} \frac{\varphi(x)}{2 \tan \frac{x}{2}}\,\mathrm{d}x.$$

In particular, we have

$$\lim_{\varepsilon \to 0} \int_{\varepsilon \leq |x| \leq \pi} \frac{\mathrm{e}^{-\mathrm{i}jx}}{2 \tan \frac{x}{2}} \frac{\mathrm{d}x}{2\pi} = \lim_{m \to +\infty} \int_{-\pi}^{\pi} D_m^*(x)\,\mathrm{e}^{-\mathrm{i}jx} \frac{\mathrm{d}x}{2\pi} = -\frac{\mathrm{i}\,\mathrm{sgn}(j)}{2}.$$

In other words, the Fourier transform of the distribution pv $\dfrac{1}{2 \tan \frac{x}{2}}$ on the torus is the function $-\frac{\mathrm{i}\,\mathrm{sgn}}{2}$.

If f is C^1, the convolution $\tilde{f} = f * \mathrm{pv} \frac{1}{2 \tan \frac{x}{2}}$ is a function and we have $\widehat{\tilde{f}}(j) = -\frac{\mathrm{i}\,\hat{f}(j)\,\mathrm{sgn}(j)}{2}$.

The function $\tilde{f}$ is called the conjugate function to f and the operation $f \mapsto \tilde{f}$ the Hilbert transform on the torus.

Due to the Parseval identity, the Hilbert transform extends as a bounded operator on $L^2(\mathbb{T})$.

In fact, it extends as a bounded operator of $L^p(\mathbb{T})$ for $1 < p < +\infty$. There are numerous proofs of this fact. In particular we could develop the Calderón–Zygmund theory in this context. We prefer to show in the next section that this is a consequence of the work done on $\mathbb{R}$.

If $f \in L^2(\mathbb{T})$ and $m, n \in \mathbb{Z}$, with $m \leq n$, set

$$S_{m,n}f(x) = \sum_{m \leq j \leq n} \hat{f}(j)\,\mathrm{e}^{\mathrm{i}jx}.$$

As previously indicated, the operators $S_{m,n}$ are closely related to the Hilbert transform. Indeed, we have

$$S_{0,+\infty}f = \frac{f + \hat{f}(0) + 2\mathrm{i}\tilde{f}}{2}.$$

It is easy to check the following formulae:

$$S_{m,+\infty}f = e^{imx}S_{0,+\infty}(e^{-imx}f),$$
$$S_{-\infty,n}f = (S_{-n,+\infty}\check{f})\check{},$$
$$S_{m,n} = S_{-\infty,n}S_{m,+\infty},$$

where $\check{f}(x) = f(-x)$.

This means that $S_{m,n}f$ is obtained by composing operator $S_{0,+\infty}$, multiplications by exponentials, and the involution $f(x) \mapsto \check{f}(x) = f(-x)$.

7.1.3 ▪ L^p-boundedness of partial sum operators

Let sinc stand for the function $\operatorname{sinc} x = \dfrac{\sin \pi x}{\pi x}$, which is the Fourier transform of $\mathbf{1}_{[-1/2,1/2]}$. This function, called *cardinal sine*, is of wide use in signal processing.

Lemma 7.1.

1. *Let f be a trigonometric polynomial, $f(x) = \sum \hat{f}(j)e^{ijx}$. Then the Fourier transform of the function $g(x) = f(2\pi x)\operatorname{sinc} x$ is*

$$\hat{g}(y) = \sum \hat{f}(j)\mathbf{1}_{[-1/2,1/2]}(y-j).$$

2. *For all $p \in (1,+\infty)$ there exists C_p such that, for all trigonometric polynomial,*

$$\frac{2}{\pi}\|f\|_{L^p(\mathbb{T})} \le \|g\|_{L^p(\mathbb{R})} \le C_p\|f\|_{L^p(\mathbb{T})}.$$

Proof. We have

$$\hat{g}(y) = \int_{\mathbb{R}} \left(\sum_{j\in\mathbb{Z}} \hat{f}(j)\,e^{2i\pi jx} \right) e^{-2i\pi xy}\operatorname{sinc} x\,dx$$
$$= \sum_{j\in\mathbb{Z}} \hat{f}(j)\mathbf{1}_{[-1/2,1/2]}(y-j),$$

$$\int_{-\infty}^{+\infty} |g(x)|^p\,dx = \sum_{n\in\mathbb{Z}} \int_{n-1/2}^{n+1/2} |g(x)|^p dx$$
$$= \sum_{n\in\mathbb{Z}} \int_{n-1/2}^{n+1/2} |f(2\pi x)|^p|\operatorname{sinc} x|^p dx$$
$$= \sum_{n\in\mathbb{Z}} \int_{-1/2}^{1/2} |f(2\pi x)|^p|\operatorname{sinc}(x+n)|^p dx.$$

So

$$\int_{-\infty}^{+\infty} |g(x)|^p\,dx \ge \inf_{|x|\le 1/2}|\operatorname{sinc} x|^p \int_{-1/2}^{1/2} |f(2\pi x)|^p dx = \left(\frac{2}{\pi}\right)^p\|f\|_{L^p(\mathbb{T})}^p$$

and

$$\int_{-\infty}^{+\infty} |g(x)|^p \, \mathrm{d}x \leq \|f\|_{L^p(\mathbb{T})}^p \left(1 + 2 \sum_{n \geq 1} \pi^{-p}(n - 1/2)^{-p} \right)$$

$$= \left(1 + 2\pi^{-p}(2^p - 1)\zeta(p) \right) \|f\|_{L^p(\mathbb{T})}^p.$$

Remark 7.2. *Lemma 7.1 is valid not only for scalar valued functions f but also for functions taking their values in a Hilbert space.*

7.1.4 ▪ Multipliers

Lemma 7.3. *Let m be a function defined on $\mathbb{Z}$. If, for some $p \in (1, +\infty)$, the function*
$$y \mapsto \sum_{j \in \mathbb{Z}} m(j) \mathbf{1}_{[-1/2, 1/2]}(y - j) \text{ is a multiplier of } \mathscr{F}\left(L^p(\mathbb{R})\right), \text{ then } m \text{ is a multiplier}$$
of $\mathscr{F}\left(L^p(\mathbb{T})\right)$.

Proof. Set $\tilde{m}(y) = \sum_{j \in \mathbb{Z}} m(j) \mathbf{1}_{[-1/2, 1/2]}(y - j)$. Consider a trigonometric polynomial f and set $g(x) = f(2\pi x) \operatorname{sinc} x$. Define Tf and $\tilde{T}g$ by the following formulae:

$$\widehat{Tf}(j) = m(j)\hat{f}(j) \qquad \text{and} \qquad \widehat{\tilde{T}g}(y) = \tilde{m}(y)\hat{g}(y).$$

It is easy to check that $\tilde{T}g(x) = Tf(2\pi x) \operatorname{sinc} x$. Then Lemma 7.3 yields

$$\|Tf\|_{L^p(\mathbb{T})} \leq \frac{\pi}{2} \|\tilde{T}g\|_{L^p(\mathbb{R})}$$
$$\leq C\|g\|_{L^p(\mathbb{R})} \leq CC_p\|f\|_{L^p(\mathbb{T})}.$$

Corollary 7.4. *The partial sum operators $S_{m,n}$ ($m \leq n$) are uniformly bounded on $L^p(\mathbb{T})$ for all $p \in (1, +\infty)$.*

Corollary 7.5 (Marcinkiewicz multiplier theorem). *Let m be a bounded function on $\mathbb{Z}$ such that*

$$\sup_{j > 0} \sum_{2^j \leq k < 2^{j+1}} \left(|m(k+1) - m(k)| + |m(-k-1) - m(-k)| \right) < +\infty.$$

Then, for all $p \in (1, +\infty)$, m is a multiplier of $\mathscr{F}\left(L^p(\mathbb{T})\right)$.

7.2 ▪ $\mathbb{Z}$

If $a = \left(a(j)\right)_{j \in \mathbb{Z}}$ is a finite sequence, we set

$$Ha(k) = \sum_{j \neq k} \frac{a(j)}{k - j}.$$

Theorem 7.6. *For all $p \in (0, +\infty)$ there exists C_p such that, for all finite sequence a, we have*

$$\|Ha\|_p \leq C_p\|a\|_p.$$

Proof. A proof was already outlined in Exercise 3.5. The proof below is another example of transference.

We begin with two remarks.

- If $\alpha > 0$, the Hilbert transform of $\mathbf{1}_{[-\alpha,\alpha]}$ is $\log \left| \dfrac{x+\alpha}{x-\alpha} \right|$.

- If $|t| \leq 1/2$, then $\left| \log \dfrac{1+t}{1-t} - 2t \right| \leq t^2$.

If k is a nonzero integer and $|x - k| \leq 1/2$, we have $\dfrac{1}{4|x|} \leq \dfrac{1}{2}$ and

$$\left| \log \frac{1 + \frac{1}{4x}}{1 - \frac{1}{4x}} - \frac{1}{2k} \right| \leq \left| \log \frac{1 + \frac{1}{4x}}{1 - \frac{1}{4x}} - \frac{1}{2x} \right| + \left| \frac{1}{2x} - \frac{1}{2k} \right|$$

$$\leq \frac{1}{16x^2} + \frac{1}{4|kx|}$$

$$\leq \frac{5}{16 \left(|k| - \frac{1}{2} \right)^2}. \tag{7.2}$$

Consider the function $f(x) = \sum_{j \in \mathbb{Z}} a(j) \mathbf{1}_{[-1/4, 1/4]}(x - j)$. Its Hilbert transform is (there is no harm using the symbol H again)

$$Hf(x) = \sum_{j \in \mathbb{Z}} a(j) \log \left| \frac{\frac{1}{4} + (x-j)}{\frac{1}{4} - (x-j)} \right|.$$

If, as previously, $|x - k| \leq 1/2$, with $k \in \mathbb{Z} \setminus \{0\}$, we have

$$Hf(x) = \sum_{j \neq k} \frac{a(j)}{2(k-j)} + g_1(x) + g_2(x), \tag{7.3}$$

where

$$g_1(x) = a(k) \log \left| \frac{\frac{1}{4} + (x-k)}{\frac{1}{4} - (x-k)} \right|,$$

$$g_2(x) = \sum_{j \neq k} a(j) \left(\log \frac{1 + \frac{1}{4(x-j)}}{1 - \frac{1}{4(x-j)}} - \frac{1}{2(k-j)} \right).$$

It results from (7.2) that

$$|g_2(x)| \leq \sum_{j \neq k} \frac{5a(j)}{4 \left(2|k-j| - 1 \right)^2} = a * b(k),$$

where $b \in \ell^1$ is so defined: $b(0) = 0$, and $b(j) = \frac{5}{4(|j|^2 - 1)}$ if $j \neq 0$.

Therefore, we have $\left(\int_{\mathbb{R}} |g_2(x)|^p \, dx \right)^{1/p} \leq \|a\|_{\ell^p} \|b\|_{\ell^1}$. We also have

$$\int_{\mathbb{R}} |g_1(x)|^p dx = \|a\|_{\ell^p}^p \int_{-1/2}^{1/2} \left| \log \frac{|1 + 4x|}{|1 - 4x|} \right|^p dx.$$

Then (7.3) yields

$$\|Ha\|_{\ell^p} \leq 2\|Hf\|_{L^p} + C_p\|a\|_{\ell^p}.$$

One concludes by using the boundedness of the Hilbert transform on L^p:

$$\|Hf\|_{L^p} \leq A_p\|f\|_{L^p} = 2^{-1/p}A_p\|a\|_{\ell^p}.$$

7.3 ▪ Some totally disconnected groups

7.3.1 ▪ The setting

Let $\mathbb{J}$ stand for either $\mathbb{Z}$ or $\mathbb{Z}_+ = \{0, 1, 2, \ldots\}$. We consider a locally compact abelian group G with a decreasing sequence $(G_j)_{j \in \mathbb{J}}$ of open compact subgroups such that $\bigcup G_j = G$ and $\bigcap G_j = \{0\}$. The index $(G_j : G_{j+1})$, which we shall denote by $\varkappa_j$, is the cardinality of the quotient group G_j/G_{j+1}.

The classes modulo the subgroups G_j generate the topology of G. To define a Haar measure on G, due to its invariance par translations, it is sufficient to set its values on each G_j. We choose as Haar measure on G the measure $\mathcal{L}_G$ which satisfies

$$\mathcal{L}_G(G_0) = 1 \quad \text{and} \quad \mathcal{L}_G(G_j) = \varkappa_j \, \mathcal{L}_G(G_{j+1}) \quad \text{for all} \quad j \in \mathbb{J}.$$

When there may be no ambiguity, we write $\mathcal{L}$ instead of $\mathcal{L}_G$. As in the previous chapters, we shall often write $\mathrm{d}x$ instead of $\mathrm{d}\mathcal{L}(x)$. We also often write $|G_j|$ or g_j instead of $\mathcal{L}(G_j)$. Also, $L^p(G)$ and $L^p(\mathcal{L}_G)$ are synonyms.

It is useful to notice that the locally constant and compactly supported functions (the simple functions) are dense in all the $L^p(G)$ $(1 \leq p < \infty)$ and in the space of continuous functions on G which vanish at ∞.

Hereafter, we suppose that there exists a bound $\varkappa$ such that, for all j, $\varkappa_j \leq \varkappa$.

Let Γ be the dual group of G. Recall that Γ is the set of continuous homomorphisms from G to the multiplicative group of complex numbers of modulus 1. These homomorphisms are called the *continuous characters* of G. Endowed with the topology of the uniform convergence on compact subsets, Γ is a locally compact group. It is naturally endowed with an increasing sequence of compact open subgroups: $\Gamma_j = \{\gamma \in \Gamma : \gamma(G_j) = \{1\}\}$.

It is easy to check that $(\Gamma_{j+1} : \Gamma_j) = (G_j : G_{j+1}) = \varkappa_j$. We choose the Haar measure on Γ, $\mathcal{L}_\Gamma$, such that $\mathcal{L}_\Gamma(\Gamma_0) = 1$ and $\mathcal{L}_\Gamma(\Gamma_{j+1}) = \varkappa_j \, \mathcal{L}_\Gamma(\Gamma_j)$. So, we have $\mathcal{L}_\Gamma(\Gamma_j)\mathcal{L}_G(G_j) = 1$.

If $f \in L^1(G)$, the Fourier transform of f is the function on Γ so defined:

$$\hat{f}(\gamma) = \int_G \overline{\gamma(x)} \, f(x) \, \mathrm{d}x.$$

It is easy to check that $\widehat{f * g} = \hat{f}\hat{g}$. Also the Fourier transform of $\mathbf{1}_{G_j}$ is $\mathcal{L}(G_j) \mathbf{1}_{\Gamma_j} = |G_j| \, \mathbf{1}_{\Gamma_j} = \mathrm{g}_j \mathbf{1}_{\Gamma_j}$.

In what follows, we use the notation $\chi_j = \mathrm{g}_j^{-1}\mathbf{1}_{G_j}$. Then $\hat{\chi}_j = \mathbf{1}_{\Gamma_j}$.

With the above choices of Haar measures the Fourier transform extends as an isometry from $L^2(G)$ to $L^2(\Gamma)$.

The inverse Fourier transform of $f \in L^1(\Gamma)$ is the following function on G:

$$\mathscr{F}^{-1}f(x) = \int_\Gamma \gamma(x)f(\gamma)\,\mathrm{d}\gamma.$$

It might be convenient to endow G and Γ with metrics. If $x \in G_j \setminus G_{j+1}$, set $|x| = \mathsf{g}_j$ and $|0| = 0$. Then $(x, y) \mapsto |x - y|$ is an ultrametric on G. The corresponding balls are just the classes of the subgroups G_j.

Similarly, for $\gamma \in \Gamma_{j+1} \setminus \Gamma_j$, set $|\gamma| = |\Gamma_j| = \mathsf{g}_j^{-1}$ and $|0| = 0$.

7.3.2 ▪ Singular integrals

Define the maximal function:

$$\mathsf{M}f(x) = \sup_{j \in \mathbb{J}} \frac{1}{|G_j|} \int_{G_j} |f(x - y)|\,\mathrm{d}y = \sup_{j \in \mathbb{J}} |f| * \chi_j(x).$$

Then, as the classes modulo the subgroups G_j have the same properties as the dyadic cubes (if two of them intersect, then one contains the other), one obtains $\mathcal{L}(\mathsf{M}f > t) \leq t^{-1} \int_{\{\mathsf{M}f>t\}} |f|\,\mathrm{d}\mathcal{L}$. This implies that for $p > 1$ one has

$$\|\mathsf{M}f\|_p \leq \frac{p}{p-1}\|f\|_p$$

(see Exercise 1.7).

Another consequence is that in this context the Lebesgue derivation theorem is valid: if f is locally integrable, for $\mathcal{L}$-almost every x in G, one has $\lim_{j \to +\infty} \frac{1}{|G_j|} \int_{G_j} |f(x - y) - f(x)|\,\mathrm{d}y = 0$.

Theorem 7.7. *Let K be a square integrable function on G such that*

a. $\|\hat{K}\|_\infty < +\infty$;

b. $B = \sup_{y \in G \setminus \{0\}} \int_{|x|>|y|} |K(x - y) - K(x)|\,\mathrm{d}x < +\infty.$

Then, if T is the operator $Tf = K * f$,

1. *there exits C depending only on κ, B, and $\|\hat{K}\|_\infty$ such that*

$$\mathcal{L}(|Tf| > t) \leq Ct^{-1}\|f\|_1$$

for all $f \in L^1(G)$ and $t > 0$;

2. *T extends as a bounded operator on $L^p(G)$ for $1 < p < \infty$; more precisely, for any $p \in (1, +\infty)$, there exists C_p depending only on κ, p, B, and $\|\hat{K}\|_\infty$ such that*

$$\|Tf\|_p \leq C_p\|f\|_p$$

for all simple function f.

Proof. We use classes of subgroups G_j instead of dyadic cubes to establish a Calderón–Zygmund lemma where 2^n is replaced by κ. The rest of the proof is identical to that of Theorem 3.5.

Obviously, Theorem 7.7 like Theorem 3.5 has a vector version.

Remark 7.8. *If, for all $j \in \mathbb{J}$, the kernel K is constant on each class of G_{j+1} contained in $G_j \setminus G_{j+1}$, then it fulfills the second hypothesis of Theorem 7.7 with $B = 0$.*

Theorem 7.9. *If f is a locally integrable function on G, set*

$$
\mathsf{S}f = \begin{cases} \left(\sum_{j\in\mathbb{Z}}|(\chi_{j+1}-\chi_j)*f|^2\right)^{\frac{1}{2}} & \text{if}\quad \mathbb{J}=\mathbb{Z}, \\[2mm] \left(|f*\chi_0|^2 + \sum_{j\geq 0}|(\chi_{j+1}-\chi_j)*f|^2\right)^{\frac{1}{2}} & \text{if}\quad \mathbb{J}=\mathbb{Z}_+. \end{cases}
$$

Then, for all $p \in (1, +\infty)$, there exists C_p such that, for all $f \in L^p(G)$, one has

$$
C_p^{-1}\|f\|_{L^p(G)} \leq \|\mathsf{S}f\|_{L^p(G)} \leq C_p\|f\|_{L^p(G)}.
$$

Proof. Let J be a finite subset of $\mathbb{J}$. Consider

$$
K = \left(\chi_{j+1}-\chi_j\right)_{j\in J}.
$$

This is a function from G to the Hilbert space $\ell^2(J)$. The kernel K is L^2, satisfies the property in Remark 7.8, and its Fourier transform is bounded (because the Fourier transform of $\chi_{j+1}-\chi_j$ is $\mathbf{1}_{\Gamma_{j+1}} - \mathbf{1}_{\Gamma_j}$). Then the vector version of Theorem 7.7 yields the inequality

$$
\left\|\left(\sum_{j\in J}|(\chi_{j+1}-\chi_j)*f|^2\right)^{1/2}\right\|_{L^p(G)} \leq C_p\|f\|_{L^p(G)}
$$

with a bound C_p which does not depend on J. Then, by monotone convergence, one gets $\|\mathsf{S}f\|_{L^p(G)} \leq C_p\|f\|_{L^p(G)}$.

But the operators $f \mapsto (\chi_{j+1}-\chi_j)*f$ are projectors, which are mutually orthogonal and span $L^2(G)$ (in case $\mathbb{J} = \mathbb{Z}_+$, one has to add the projector $f \mapsto f*\chi_0$). So Lemma 4.8 yields the converse inequality.

Exercise 7.2 yields another proof of this theorem.

Corollary 7.10. *If $f \in L^p(G)$, set $f_j = \chi_j * f$. Let $(\psi_j)_{j\in\mathbb{J}}$ be a sequence of functions such that*

1. *$\sup \|\psi_j\|_{L^\infty(G)} < +\infty$;*

2. *for all $j \in \mathbb{J}$, ψ_j is constant on the classes modulo G_j.*

Then, for all $f \in L^p(G)$, the series $\sum_{j\in\mathbb{J}}(f_{j+1} - f_j)\psi_j$ converges in $L^p(G)$ and

$$
\left\|\sum_{j\in\mathbb{J}}(f_{j+1} - f_j)\psi_j\right\|_{L^p(G)} \leq C_p^2\|f\|_{L^p(G)}\sup_{j\in\mathbb{J}}\|\psi_j\|_{L^\infty(G)}.
$$

Proof. Set $M = \sup\|\psi_j\|_\infty$. We have

$$
\left((f_{j+1}-f_j)\psi_j\right)*\chi_k = \begin{cases} 0 & \text{if}\quad k\leq j, \\ (f_{j+1}-f_j)\psi_j & \text{if}\quad k > j. \end{cases}
$$

So

$$
\Big((f_{j+1} - f_j)\psi_j\Big) * (\chi_{k+1} - \chi_k) = \begin{cases} 0 & \text{if } k \neq j, \\ (f_{j+1} - f_j)\psi_j & \text{if } k = j. \end{cases} \tag{7.4}
$$

Let $j_0 < j_1$ be two elements of $\mathbb{J}$. Consider $g = \sum_{j=j_0}^{j_1} (f_{j+1} - f_j)\psi_j$. Due to (7.4), $g * (\chi_{j+1} - \chi_j) = (f_{j+1} - f_j)\psi_j$ for $j_0 \leq j \leq j_1$. Then, by applying Theorem 7.9 twice, we get

$$
\|g\|_p \leq C_p \left\| \left(\sum_{j=j_0}^{j_1} |(f_{j+1} - f_j)\psi_j|^2 \right)^{\frac{1}{2}} \right\|_p
$$

$$
\leq M C_p \left\| \left(\sum_{j=j_0}^{j_1} |f_{j+1} - f_j|^2 \right)^{\frac{1}{2}} \right\|_p
$$

$$
\leq M C_p^2 \|f_{p_1+1} - f_{p_0}\|_p.
$$

By the Cauchy criterion, the series $\sum_{j \in \mathbb{J}} (f_{j+1} - f_j)\,\psi_j$ converges in $L^p(G)$ and the L^p norm of its sum is dominated by $2 M C_p^2 \|f\|_p$.

Remark 7.11 (probabilistic interpretation). *Now we deal with the case $\mathbb{J} = \mathbb{Z}_+$. Then $G = G_0$. Endowed with its Haar measure, G is a probability space. Let $\mathscr{A}_j$ be the Boolean algebra generated by the classes of the subgroup G_j. Then $\mathbf{g}_j^{-1} f * \mathbf{1}_{G_j}$ is the conditional expectation $\mathbb{E}(f \mid \mathscr{A}_j)$.*

Consider a martingale $(f_n)_{n \geq 0}$ adapted to the filtration $(\mathscr{A}_n)$, and form $S = \left(\sum_{n \geq 0} |f_{n+1} - f_n|^2 \right)^{\frac{1}{2}}$. Then Theorem 7.9 says that, for $1 < p < \infty$, this martingale is bounded in L^p if and only if S is in L^p.

Now, we are given a sequence $(\psi_n)_{n \geq 0}$ of uniformly bounded functions such that ψ_n is $\mathscr{A}_n$-measurable. We have

$$
\mathbb{E}\big((f_{n+1} - f_n)\psi_n \mid \mathscr{A}_n\big) = \psi_n \, \mathbb{E}\big(f_{n+1} - f_n \mid \mathscr{A}_n\big) = 0.
$$

Therefore $\sum_{j=0}^{n} (f_{j+1} - f_j)\psi_j$ is a martingale, bounded in L^p if the original martingale is bounded in L^p.

7.3.3 ▪ More singular integrals

Lemma 7.12. *Let K be a function on $G \setminus \{0\}$ such that*

1. $\displaystyle \sup_{j \in \mathbb{J}} \int_{G_j \setminus G_{j+1}} |K(x)| \, \mathrm{d}x < +\infty$;

2. $\displaystyle \sup_{y \neq 0} \int_{|x| > |y|} |K(x - y) - K(x)| \, \mathrm{d}x < +\infty$;

3. *for all $j \in \mathbb{J}$,* $\displaystyle\int_{G_j \setminus G_{j+1}} K(x)\,\mathrm{d}x = 0.$

Then the Fourier transform of K is bounded.

Proof. Consider $m \in \mathbb{J}$ and $\gamma \in \Gamma_{m+1} \setminus \Gamma_m$. Then

$$\hat{K}(\gamma) = \int_{G \setminus G_{m+1}} \overline{\gamma(x)}\, K(x)\,\mathrm{d}x,$$

because of the third hypothesis.

Take $y \in G_m \setminus G_{m+1}$ such that $\gamma(y) \neq 1$; this is possible because $\gamma \notin \Gamma_m$. Let k be the smallest positive integer such that $ky \in G_{m+1}$. Such a k exists and is less than or equal to $(G_m : G_{m+1})$: k is the order of the class of y modulo G_{m+1} in the quotient group G_m/G_{m+1}. Set $y_0 = ky$ and, for $1 \le l < k$, set $y_l = ly$. Since $\gamma(y) \neq 1$ is a kth root of 1, we have $\sum_{l=0}^{k-1} \gamma(ly) = 0$, i.e., $\sum_{l=1}^{k-1} \gamma(ly) = -1$.

Then

$$\hat{K}(\gamma) = \frac{1}{k} \sum_{l=0}^{k-1} \gamma(y_l) \int_{G \setminus G_{m+1}} \overline{\gamma(x + y_l)}\, K(x)\,\mathrm{d}x$$

$$= \frac{1}{k} \sum_{l=1}^{k-1} \gamma(y_l) \left(\int_{G \setminus G_{m+1}} \overline{\gamma(x + y_l)}\, K(x)\,\mathrm{d}x - \int_{G \setminus G_{m+1}} \overline{\gamma(x + y_0)}\, K(x)\,\mathrm{d}x \right)$$

$$= \frac{1}{k} \sum_{l=1}^{k-1} \gamma(y_l) \left(\int_{-y_l + {}^c G_{m+1}} \overline{\gamma(x)}\, K(x - y_l)\,\mathrm{d}x - \int_{-y_0 + {}^c G_{m+1}} \overline{\gamma(x)}\, K(x - y_0)\,\mathrm{d}x \right)$$

$$= \frac{1}{k} \sum_{l=1}^{k-1} \gamma(y_l) \left(\int_{-y_l + {}^c G_{m+1}} \overline{\gamma(x)}\, K(x - y_l)\,\mathrm{d}x - \int_{{}^c G_{m+1}} \overline{\gamma(x)}\, K(x)\,\mathrm{d}x \right).$$

Set $\displaystyle A_l = \int_{-y_l + {}^c G_{m+1}} \overline{\gamma(x)}\, K(x - y_l)\,\mathrm{d}x - \int_{{}^c G_{m+1}} \overline{\gamma(x)}\, K(x)\,\mathrm{d}x.$

Since ${}^c G_{m+1} = \left(-y_l + G_m \setminus G_{m+1} \right) \cup \left(\bigcup_{n<m} G_n \setminus G_{n+1} \right)$, we have

$$A_l = \int_{{}^c G_m} \overline{\gamma(x)}\big(K(x - y_l) - K(x) \big)\,\mathrm{d}x$$

$$+ \int_{G_m \setminus G_{m+1} - y_l} \overline{\gamma(x)}\big(K(x - y_l) \big)\,\mathrm{d}x - \int_{G_m \setminus G_{m+1}} \overline{\gamma(x)}\big(K(x) \big)\,\mathrm{d}x.$$

So we have

$$|A_l| \le \int_{|x|>|y_l|} \big| K(x - y_l) - K(x) \big|\,\mathrm{d}x + 2 \int_{G_m \setminus G_{m+1}} |K(x)|\,\mathrm{d}x$$

$$\le \sup_{z \neq 0} \int_{|x|>|z|} \big| K(x - z) - K(x) \big|\,\mathrm{d}x + 2 \sup_{j \in \mathbb{J}} \int_{G_j \setminus G_{j+1}} |K(x)|\,\mathrm{d}x.$$

Lemma 7.13. *Let Ω be a bounded function on $G\{0\}$. For $t > 0$, set*

$$\omega(t) = \sup_{j \in \mathbb{J}} \sup_{|x| = \mathsf{g}_j} \sup_{|y| \le t|x|} \big| \Omega(x - y) - \Omega(x) \big|.$$

Assume

1. $\displaystyle\sup_{j\in\mathbb{J}}\sum_{\substack{k\leq j\\ k\in\mathbb{J}}}\omega\left(g_{j+1}/g_k\right)<+\infty;$

2. *for all* $j\in\mathbb{J},\ \displaystyle\int_{|x|=g_j}\Omega(x)\,\mathrm{d}x=0.$

Then the kernel $K(x)=\Omega(x)/|x|$ *fulfills the hypotheses of Lemma 7.12.*

Proof. The proof is left as an exercise.

Example 7.14. *Consider* $G=\mathbb{Z}_p$ *the ring of p-adic integers. Let* Ω *be a function on* $\mathbb{Z}_p\setminus p\mathbb{Z}_p$ *satisfying the Dini condition and of mean* 0. *Let* $(a_n)_{n\geq 0}$ *be a sequence of units of* $\mathbb{Z}_p$. *Then the kernel* $\dfrac{\Omega(a_{v_p(x)}|x|^{-1}x)}{|x|}$, *where* $v_p(x)$ *is the p-adic valuation of* x, *defines a singular integral operator which is bounded on* $L^r(\mathbb{Z}_p)$ *for all* $r\in(0,+\infty)$.

7.3.4 ▪ More on martingales

Consider a tree $\mathscr{T}$ with root ϵ. For two nodes u and v, let $\delta(u,v)$ stand for the geodesic distance in $\mathscr{T}$ between u and v (this is the minimal length of a path from u to v). Let $\mathscr{T}_n$ stand for the set of nodes which are at distance n from the root ϵ.

For a node v of $\mathscr{T}$, let $\overline{v}$ stand for the set of its offsprings, and $\dot{v}$, if $v\neq\epsilon$, stand for its parent:

$$\overline{v}=\big\{w\ :\ \delta(v,w)=1,\text{ and }\delta(\epsilon,w)=\delta(\epsilon,v)+1\big\},$$

$\dot{v}$ is the node w such that $v\in\overline{w}$.

This tree is endowed with a positive function λ satisfying the following three conditions:

1. $\lambda(\epsilon)=1.$

2. For all $v\in\mathscr{T}$, $\displaystyle\lambda(v)=\sum_{w\in\overline{v}}\lambda(w).$

3. $\varkappa=\displaystyle\sup_{v\in\mathscr{T}}\sup_{w\in\overline{v}}\frac{\lambda(v)}{\lambda(w)}$ is finite.

Under these conditions, it is obvious that $\operatorname{card}\overline{v}\leq\varkappa$ for all v.

If v and w are two nodes of $\mathscr{T}$, $v\wedge w$ stands for their first common ancestor. Then $\mathrm{d}(v,w)=\lambda(v\wedge w)$ defines an ultrametric on $\mathscr{T}$. The completion $\overline{\mathscr{T}}$ is the union of $\mathscr{T}$ and $\mathscr{T}^*$, a compact totally disconnected set, with no isolated points since λ is positive. The set $\mathscr{T}^*$ can be identified with the branches issued from ϵ. For $v\in\mathscr{T}$, let $[v]$ stand for the subset of $\mathscr{T}^*$ whose elements are the branches going through the node v. These subsets $[v]$ are called *cylinders* and are the balls for the distance, still denoted by d, induced by d on $\mathscr{T}^*$. The function λ defines a probability measure, also denoted by λ, on $\mathscr{T}^*$ by setting

$$\lambda([v])=\lambda(v)$$

for all $v\in\mathscr{T}$ and using the Carathéodory extension theorem.

If K is a function on $\mathscr{T}^* \times \mathscr{T}^*$ such that $K(x, \grave{)} \in L^2(\lambda)$ for almost all x, for $\in L^2\lambda$, we set

$$Kf(x) = \int K(x,y)f(y)\,\mathrm{d}\lambda(y).$$

Theorem 7.15. *Let K be a kernel such that*

1. *the associated operator $T : f \mapsto Kf$ is bounded on $L^2(\lambda)$;*

2. $B = \sup\limits_{v \in \mathscr{T}} \sup\limits_{y,z \in [v]} \int_{x \notin [v]} |K(x,y) - K(x,z)|\,\mathrm{d}\lambda(x) < +\infty.$

Then, for all $p \in (1, +\infty)$, the operator T extends as a bounded operator on $L^p(\lambda)$ and its norm is bounded by a constant depending only on $\varkappa$, p, B.

Proof. The cylinders play the role of dyadic cubes in this context: there are analogues of Hardy–Littlewood lemma and of Calderón–Zygmund decomposition. See also Exercise 3.4.

Let $\mathscr{A}_n$ be the σ-field generated by $\{[v] ; v \in \mathscr{T}_n\}$. If $X \in L^1(\lambda)$, the conditional expectation $\mathbb{E}(X \mid \mathscr{A}_n)$ is given by the kernel

$$L_n(x,y) = \sum_{v \in \mathbb{T}_n} \frac{1}{\lambda([v])} \mathbf{1}_{[v]}(y)\mathbf{1}_{[v]}(x),$$

so that

$$\mathbb{E}(X \mid \mathscr{A}_n) = L_n X.$$

Remark 7.16. *The following property is easily checked: if $n \geq 0$, $w \in \mathscr{T}$, $x \notin [w]$, $y \in [w]$, and $z \in [w]$, then $L_n(x,y) = L_n(x,z)$.*

Theorem 7.17. *Let $(X_n)_{n\geq 0}$ be a martingale on the filtration $(\mathscr{A}_n)_{n\geq 0}$. Set*

$$S = \left(|X_0|^2 + \sum_{n\geq 1} |X_n - X_{n-1}|^2\right)^{1/2}.$$

Then, for all $p \in (1, +\infty)$, this martingale is bounded in L^p if and only if $\|S\|_p < +\infty$.

Proof. The kernels L_0 and $(L_{j-1} - L_j)_{n\geq 1}$ define a sequence of mutually orthogonal projectors in $L^2(\lambda)$ which add up to identity.

Set $K_n = \varepsilon_0 L_0 + \sum_{j=1}^n \varepsilon_j (L_{j-1} - L_j)$, where the ε_j are the Rademacher functions. Then, regardless of what values ε_j are, the kernel K_n defines an operator of norm 1 on $L^2(\lambda)$. In view of Remark 7.16, this kernel fulfills the hypotheses of Theorem 7.15 with bounds independent of n and of the values of the ε_j. Therefore, if $p \in (1/ + \infty)$, there exists $\gamma_p > 0$ such that, for all martingales,

$$\left\|\varepsilon_0 X_0 + \sum_{j=1}^n \varepsilon_j(X_{j-1} - X_j)\right\|_p \leq \gamma_p \|X_n\|_p.$$

Then Theorem A.4 yields

$$\left\| \left(|X_0|^2 + \sum_{j=1}^{n} |X_{j-1} - X_j|^2 \right)^{\frac{1}{2}} \right\|_p \le \gamma_p C_p \|X_n\|_p. \tag{7.5}$$

This proves that if the martingale is bounded in L^p, S is in L^p.

Then, by revisiting the proof of Lemma 4.8, inequality (7.5) gives

$$\|X_n\|_p \le \gamma_{p'} C_{p'} \left\| \left(|X_0|^2 + \sum_{j=1}^{n} |X_{j-1} - X_j|^2 \right)^{\frac{1}{2}} \right\|_p \le \gamma_{p'} C_{p'} \|S\|_p.$$

This ends the proof.

Corollary 7.18. *Let $(X_n)_{n\ge 0}$ be a martingale bounded for some $p \in (1, +\infty)$, and let $(Y_n)_{n\ge 0}$ be a sequence of random variables such that $\sup \|Y_n\|_\infty < +\infty$ and such that, for all n, Y_n is measurable with respect to $\mathscr{A}_n$. Then $\left(\sum_{j=1}^{n} Y_{j-1}(X_j - X_{j-1}) \right)_{n\ge 1}$ is a martingale bounded in L^p.*

7.4 ▪ Exercises

Exercise 7.1. *This exercise is another example of transference: it shows that the L^p-boundedness of the Hilbert transform on the torus implies the L^p-boundedness of the Hilbert on the line.*

Let p be a number such that $1 < p < +\infty$. We assume that there exists a constant C_p such that, for all φ, C^1 on the torus $\mathbb{T} = \mathbb{R}/2\pi\mathbb{Z}$, we have $\|H\varphi\|_p \le C_p\|\varphi\|_p$, where $H\varphi = \varphi * \mathrm{pv}\, \frac{1}{2\tan t/2}$.

Let f be a C^1 function on $\mathbb{R}$ which vanishes outside the interval $[-R, R]$.

1. *Observe that $\displaystyle \sup_{|x|<\pi/2} \left| \frac{1}{x} - \frac{1}{\tan x} \right| < +\infty.$*

2. *For $\varepsilon \le \pi/R$, consider the 2π-periodic function φ_ε such that $\varphi_\varepsilon(t) = f(t/\varepsilon)$ for $|t| \le \pi$. Let $\Phi_\varepsilon = H\varphi_\varepsilon$ be the Hilbert transform of φ_ε on the torus. Set*

$$F_\varepsilon(x) = \begin{cases} \Phi_\varepsilon(\varepsilon x) & \text{if } |x| \le \frac{\pi}{\varepsilon}, \\ 0 & \text{otherwise}. \end{cases}$$

Show that $\|F_\varepsilon\|_{L^p(\mathbb{R})} \le C_p\|f\|_{L^p(\mathbb{R})}.$

3. *Set $\displaystyle \Psi_\varepsilon(x) = \int_{-R}^{R} f(y) \left(\frac{\varepsilon}{2\tan \frac{\varepsilon(x-y)}{2}} - \frac{1}{x-y} \right) \mathrm{d}y.$*

Show that $\displaystyle \lim_{\varepsilon \searrow 0} \|f\|_p^{-p} \int_{-R/\varepsilon}^{R/\varepsilon} |\Psi_\varepsilon(x)|^p \mathrm{d}x = 0.$

4. *Show that, for $|x| \leq R/\varepsilon$, $F_\varepsilon(x) = \mathrm{pv} \int_{-R}^{R} f(y) \dfrac{\varepsilon}{2 \tan \frac{\varepsilon(x-y)}{2}} \, \mathrm{d}y.$*

5. *Conclude that $\left\| f * \mathrm{pv} \dfrac{1}{x} \right\|_{L^p(\mathbb{R})} \leq C_p \|f\|_{L^p(\mathbb{R})}.$*

Exercise 7.2 (an alternate proof of Theorem 7.9). *Let J be a finite subset of $\mathbb{J}$. Set $K_J = \sum_{j \in J} \varepsilon_j (\chi_{j+1} - \chi_j)$, where the ε_j are the Rademacher functions.*

1. *Show that, regardless of what the values of the ε_j are, $\left\| \widehat{K_J} \right\|_\infty = 1$.*

2. *Show that $\displaystyle \sup_{y \in G \setminus \{0\}} \int_{|x| > |y|} |K(x - y) - K(x)| \, \mathrm{d}x < +\infty$ (use Remark 7.8).*

3. *Show that, for all $p \in (1, +\infty)$, there exists C_p, a constant independent of J, such that*
$$\|K_J * f\|_p \leq C_p \|f\|_p \quad \text{for all } f.$$

4. *Conclude by using Khintchine inequalities (Theorem A.4).*

Exercise 7.3. *We consider a group G as in Section 7.3. For each $j \in \mathbb{J}$, we choose $\varkappa_j - 1$ points $\{y_{j,k}\}_{1 \leq k < \varkappa_j}$, one in each class of G_{j+1}, different from G_{j+1}, contained in G_j. Also, we are given bounded numbers $\{a_{j,k}\}_{j \in \mathbb{J}, 1 \leq k < \varkappa_j}$ such that, for all j,*
$$\sum_{k=1}^{\varkappa_j - 1} a_{j,k} = 0,$$
and set
$$K = \sum_{j \in \mathbb{J}} \frac{1}{\mathsf{g}_j} \sum_{k=1}^{\varkappa_j - 1} a_{j,k} \mathbf{1}_{y_{j,k} + G_{j+1}}.$$

1. *Show that K defines a Calderón–Zygmund convolution operator.*

2. *Let m be the multiplier associated with K. Show that m has the form*
$$\mathsf{m} = \sum_{j \in \mathbb{J}} \sum_{k=1}^{\varkappa_j - 1} b_{j,k} \mathbf{1}_{\xi_{j,k} + \Gamma_j},$$
where, for all j, $\sum_k b_{j,k} = 0$, and $\left\{ \xi_{j,k} + \Gamma_j \right\}_{1 \leq k < \varkappa_j}$ are the classes of Γ_j contained in $\Gamma_{j+1} \setminus \Gamma_j$.

Exercise 7.4. *We consider a group G as in Section 7.3. Let Ω be a bounded function on G with the following property: for some $k \geq 1$, for all $j \in \mathbb{J}$, Ω is constant one the classes of G_{j+k} contained in $G_j \setminus G_{j+1}$. Then we consider the kernel K so defined: if $x \in G_j \setminus G_{j+1}$, $K(x) = \Omega(x)/\mathsf{g}_j$.*

  Prove that the kernel K defines an operator which is bounded on all L^p for $1 < p < +\infty$.

Chapter 8

Interpolation

8.1 ▪ Real methods

8.1.1 ▪ The Marcinkiewicz interpolation theorem

Let $(X, \mathcal{A}, \mu)$ and $(Y, \mathcal{B}, \nu)$ be two σ-finite measure spaces.

Let $\mathcal{E}$ stand for a space of μ-measurable complex-valued functions such that if $f \in \mathcal{E}$, then the functions $f\mathbf{1}_{\{|f|>t\}}$ and $f\mathbf{1}_{\{|f|\leq t\}}$ also are in $\mathcal{E}$, for any $t > 0$.

Let T be an operator defined on $\mathcal{E}$ whose values are ν-measurable functions and which is subadditive, i.e., $|T(f + g)| \leq |T(f)| + |T(g)|$ for all f and g.

If $p \in]0, +\infty]$ and $q \in]0, +\infty[$, one says that T is of weak type (p, q) if there exists a constant M such that, for all $f \in \mathcal{E}$ and for all $t > 0$,

$$\nu\big(\{|Tf| > t\}\big) \leq \big[t^{-1}M\, \|f\|_{L^p(\mu)}\big]^q.$$

One says that T is of weak type (p, ∞) if there exists M such that, for all $f \in \mathcal{E}$, one has $\|Tf\|_{L^\infty(\nu)} \leq M\, \|f\|_{L^p(\mu)}$ (so weak and strong types (p, ∞) are the same).

Theorem 8.1. *Let p_0, p_1, q_0, and q_1 be numbers such that $0 < p_0 \leq q_0 \leq \infty$, $0 < p_1 \leq q_1 \leq \infty$, and $q_0 \neq q_1$. Assume T is subadditive and of weak type (p_0, q_0) and (p_1, q_1) with corresponding bounds M_0 and M_1. Then if $\tau \in (0, 1)$ and if one defines p and q by the relations*

$$\frac{1}{p} = \frac{1-\tau}{p_0} + \frac{\tau}{p_1} \qquad and \qquad \frac{1}{q} = \frac{1-\tau}{q_0} + \frac{\tau}{q_1},$$

then, for all $f \in \mathcal{E}$,

$$\|Tf\|_{L^q(\nu)} \leq K\, M_0^{1-\tau} M_1^\tau\, \|f\|_{L^p(\mu)},$$

where K depends only on p_0, p_1, q_0, q_1, and τ.

Proof. We shall give the proof only in the case when $p_0 \neq p_1$ and q_0 and q_1 are finite. One may suppose $p_0 < p_1$.

Suppose first that $M_0 = M_1 = 1$.

Take f such that $\|f\|_{L^p(\mu)} > 0$. Let α and σ be two numbers, to be chosen later on, such that $\alpha > 0$ and $\sigma(q_1 - q_0) > 0$. Let f_0 be the so defined function; $f_0 = 0$ if

$|f| \leq \alpha t^{\sigma}$, otherwise $f_0 = f$. Set $f_1 = f - f_0$. Since

$$\{|Tf| > t\} \subset \left\{|Tf_0| > \frac{t}{2}\right\} \cup \left\{|Tf_1| > \frac{t}{2}\right\},$$

and taking into account the hypotheses on T, one has

$$\varrho_{Tf}(t) \leq \left[\frac{2}{t}\left(\int_{\{|f|>\alpha t^{\sigma}\}} |f|^{p_0} d\mu\right)^{\frac{1}{p_0}}\right]^{q_0} + \left[\frac{2}{t}\left(\int_{\{|f|\leq\alpha t^{\sigma}\}} |f|^{p_1} d\mu\right)^{\frac{1}{p_1}}\right]^{q_1},$$

hence

$$\|Tf\|_q^q \leq q\, 2^{q_0} \int_0^{+\infty} \left(\int_{|f|>\alpha t^{\sigma}} |f|^{p_0} d\mu\right)^{\frac{q_0}{p_0}} t^{q-q_0-1} dt$$

$$+ q\, 2^{q_1} \int_0^{+\infty} \left(\int_{|f|\leq\alpha t^{\sigma}} |f|^{p_1} d\mu\right)^{\frac{q_1}{p_1}} t^{q-q_1-1} dt.$$

By using the generalized Minkowski inequality (see Exercise A.4), one gets

$$\|Tf\|_q^q \leq q\, 2^{q_0} \left[\int \left(\int_0^{+\infty} \mathbf{1}_{\{|f|>\alpha t^{\sigma}\}} |f|^{q_0} t^{q-q_0-1} dt\right)^{\frac{p_0}{q_0}} d\mu\right]^{\frac{q_0}{p_0}}$$

$$+ q\, 2^{q_1} \left[\int \left(\int_0^{+\infty} \mathbf{1}_{\{|f|\leq\alpha t^{\sigma}\}} |f|^{q_1} t^{q-q_1-1} dt\right)^{\frac{p_1}{q_1}} d\mu\right]^{\frac{q_1}{p_1}},$$

whence

$$\|Tf\|_q^q \leq q\, 2^{q_0} \alpha^{-\frac{q-q_0}{\sigma}} \left[\frac{1}{|q-q_0|} \int |f|^{p_0 + \frac{p_0}{q_0}\frac{q-q_0}{\sigma}} d\mu\right]^{\frac{q_0}{p_0}}$$

$$+ q\, 2^{q_1} \alpha^{-\frac{q-q_1}{\sigma}} \left[\frac{1}{|q-q_1|} \int |f|^{p_1 + \frac{p_1}{q_1}\frac{q-q_1}{\sigma}} d\mu\right]^{\frac{q_1}{p_1}}.$$

Choose σ so that $p_0 + \frac{p_0(q-q_0)}{\sigma q_0} = p_1 + \frac{p_1(q-q_1)}{\sigma q_1} = p$. Due to the definition of p and q, this is possible. One gets

$$\frac{1}{\sigma} = \frac{p}{q}\frac{\frac{1}{p_0}-\frac{1}{p}}{\frac{1}{q_0}-\frac{1}{q}} = \frac{p}{q}\frac{\frac{1}{p_1}-\frac{1}{p}}{\frac{1}{q_1}-\frac{1}{q}} = \frac{p}{q}\frac{\frac{1}{p_0}-\frac{1}{p_1}}{\frac{1}{q_0}-\frac{1}{q_1}}.$$

It is to be noticed that σ has the right sign. One gets

$$\|Tf\|_{L^q(\nu)}^q \leq q\, 2^{q_0} \alpha^{-\frac{q-q_0}{\sigma}} |q-q_0|^{-\frac{q_0}{p_0}} \|f\|_p^{\frac{pq_0}{p_0}} + q\, 2^{q_1} \alpha^{-\frac{q-q_1}{\sigma}} |q-q_1|^{-\frac{q_1}{p_1}} \|f\|_p^{\frac{pq_1}{p_1}}$$

$$= q\, 2^{q_0} \alpha^{-\frac{q_0(p-p_0)}{p_0}} |q-q_0|^{-\frac{q_0}{p_0}} \|f\|_p^{\frac{pq_0}{p_0}}$$

$$+ q\, 2^{q_1} \alpha^{-\frac{q_1(p-p_1)}{p_1}} |q-q_1|^{-\frac{q_1}{p_1}} \|f\|_p^{\frac{pq_1}{p_1}}.$$

If we take $\alpha = \|f\|_p^{\beta}$ with $\beta = q\dfrac{\frac{1}{p_1 q_0}-\frac{1}{p_0 q_1}}{\frac{1}{p_1}-\frac{1}{p_0}}$, the last inequality becomes

$$\|Tf\|_{L^q(\nu)}^q \leq \left(q\, 2^{q_0}|q-q_0|^{-\frac{q_0}{p_0}} + q\, 2^{q_1}|q-q_1|^{-\frac{q_1}{p_1}}\right) \|f\|_p^q.$$

Let $K = \left(q\, 2^{q_0} |q - q_0|^{-\frac{q_0}{p_0}} + q\, 2^{q_1} |q - q_1|^{-\frac{q_1}{p_1}} \right)^{1/q}$.

Now, we wish to drop the extra assumptions on M_0 and M_1. Let a and b be two positive numbers which will be fixed later on. Let ν' stand for the measure $b\nu$ and consider the operator $T' = aT$.

One has $\|T'f\|_{L^q(\nu')} = a b^{\frac{1}{q}} \|Tf\|_{L^p(\mu)}$ and, for $i = 0, 1$,

$$\nu' \{T'f > t\} \le b \left(\frac{a}{t}\, M_i \|f\|_{L^{p_i}(\mu)} \right)^{q_i}.$$

Choose a and b so that $a b^{\frac{1}{q_i}} M_i = 1$ for $i = 0, 1$. This implies the relation $a b^{\frac{1}{q}} M_0^{1-\tau} M_1^{\tau} = 1$. Due to the previous discussion, one has $\|T'f\|_{L^q(\nu')} \le K \|f\|_{L^p(\mu)}$, i.e.,

$$\|Tf\|_{L^q(\nu)} \le K\, a^{-1} b^{-\frac{1}{q}} \|f\|_{L^p(\mu)} = K\, M_0^{1-\tau} M_1^{\tau} \|f\|_{L^p(\mu)}.$$

The case when q_0 or q_1 is infinite requires only minor modifications: σ is still defined by the above formula; one considers the function f_0, which is equal to zero if $|f| < (A t)^{\sigma}$ and equal to f otherwise, where A is chosen so that one of the sets $\left\{ |Tf_0| > \frac{t}{2} \right\}$ or $\left\{ |Tf_1| > \frac{t}{2} \right\}$ is of zero ν-measure (as we did when we studied the Hardy–Littlewood maximal operator).

At last, it is worth noticing that one may only assume that T is quasi-subadditive, i.e., there exists a constant κ such that $|T(f + g)| \le \kappa(|Tf| + |Tg|)$.

8.1.2 ▪ An application

Lemma 8.2. *Let p be a number larger than 1, and let K be a function on $\mathbb{R}^n$ such that $\varrho_K(t) \le t^{-p}$. Then, for all $q \in [1, p')$, there exists a constant C such that, for all $f \in L^q$, one has*

$$\varrho_{K*f}(t) \le C \left(\frac{\|f\|_q}{t} \right)^r,$$

where $\dfrac{1}{r} = \dfrac{1}{p} + \dfrac{1}{q} - 1.$

Proof. Let α be a positive number and set

$$K_1 = K\, \mathbf{1}_{\{|K| > \alpha\}} \quad \text{and} \quad K_2 = K\, \mathbf{1}_{\{|K| \le \alpha\}}.$$

We have

$$\varrho_{K_1}(t) = \begin{cases} \varrho_K(t) & \text{if } t \ge \alpha, \\ \varrho_K(\alpha) & \text{if } 0 < t < \alpha \end{cases} \quad \text{and} \quad \varrho_{K_2}(t) \le \begin{cases} 0 & \text{if } t \ge \alpha, \\ \varrho_K(t) & \text{if } 0 < t < \alpha. \end{cases}$$

We may suppose that $\|f\|_q = 1$.

We have $\|K_1\|_1 \le \alpha^{1-p} + \displaystyle\int_{\alpha}^{+\infty} t^{-p}\, \mathrm{d}t = \dfrac{p}{p-1}\, \alpha^{1-p}$ and, if $q > 1$,

$$\|K_2\|_{q'}^{q'} \le q' \int_0^{\alpha} t^{q'-p-1}\, \mathrm{d}t = \frac{q'}{q' - p}\, \alpha^{q'-p}.$$

Therefore, $\|K_1 * f\|_q \leq \frac{p}{p-1}\alpha^{1-p}$ and

$$\sup |K_2 * f| \leq \left(\frac{q'}{q'-p}\right)^{\frac{1}{q'}} \alpha^{\frac{q'-p}{q'}} = \left(\frac{r}{p}\right)^{\frac{1}{q'}} \alpha^{\frac{p}{r}} \leq \alpha^{\frac{p}{r}}.$$

If $q = 1$, we also have $\sup |K_2 * f| \leq \alpha \ (= \alpha^{\frac{p}{r}})$.
  So,

$$\varrho_{K*f}\left(2\alpha^{\frac{p}{r}}\right) \leq \varrho_{K_1*f}\left(\alpha^{\frac{p}{r}}\right) + \varrho_{K_2*f}\left(\alpha^{\frac{p}{r}}\right) \leq \left(\frac{\|K_1 * f\|_q}{\alpha^{\frac{p}{r}}}\right)^q = \left(\frac{p}{p-1}\right)^q \alpha^{-p},$$

which means

$$\varrho_{K*f}(t) \leq \left(\frac{p}{p-1}\right)^q \left(\frac{t}{2}\right)^{-r}.$$

Theorem 8.3. *Let p be a number larger than 1 and let $K \in L_w^p(\mathbb{R}^n)$. Then*

1. *the mapping $f \mapsto K * f$ is of weak type $(1, p)$;*

2. *if $1 < q < p'$, the mapping $f \mapsto K * f$ is bounded from $L^q(\mathbb{R}^n)$ to $L^r(\mathbb{R}^n)$, where $\frac{1}{r} = \frac{1}{p} + \frac{1}{q} - 1$.*

Proof. Lemma 8.2 says that this mapping is of weak type (q, r) for all $q \in [1, p')$, so by using the Marcinkiewicz interpolation theorem we conclude that it is of strong type (q, r) for $1 < q < p'$.

Corollary 8.4. *For $0 < \alpha < n$, consider the function*

$$K_\alpha(x) = \frac{\pi^{-\frac{n}{2}} 2^{-\alpha}\Gamma\left(\frac{n-\alpha}{2}\right)}{\Gamma\left(\frac{\alpha}{2}\right)} |x|^{\alpha-n}$$

and the convolution operator $I_\alpha f = K_\alpha * f$. Then

1. *I_α is of weak type $\left(1, \frac{n}{n-\alpha}\right)$;*

2. *if $1 < p < q < \infty$ and $q^{-1} = p^{-1} - \frac{\alpha}{n}$, then I_α is bounded from L^p to L^q.*

Proof. Since $K_\alpha \in L_w^{\frac{n}{n-\alpha}}$ this results from the preceding theorem.

  $I_\alpha f$ is called the Riesz potential of f of order α. We already computed the Fourier transform of K_α (Remark 5.13). It results that $I_\alpha f = (-\Delta)^{-\alpha/2} f$, where Δ is the Laplacian operator. In other words, I_α is a fractional integration operator.

8.1.3 ▪ Real interpolation with BMO

Theorem 8.5. *Let T be a linear operator defined on simple functions from $\mathbb{R}^n$ to measurable functions on $\mathbb{R}^n$. We suppose that T is bounded on $L^2(\mathbb{R}^n)$ and bounded from $L^\infty(\mathbb{R}^n)$ to $\mathrm{BMO}(\mathbb{R}^n)$. Then, for any $p \in [2, +\infty)$, T is bounded on $L^p(\mathbb{R}^n)$,*

Proof. Consider the operator $\widetilde{T}$ so defined: $\widetilde{T}f = (Tf)_{\mathcal{D}}^{\sharp}$. It is sublinear and bounded on L^2 as well as on L^{∞}, so, due to the Marcinkiewicz interpolation theorem, it is bounded on L^p for $2 \leq p < +\infty$: there exists C_p such that, for any simple f, we have $\|(Tf)_{\mathcal{D}}^{\sharp}\|_p \leq C_p\|f\|_p$. But as a simple function is in L^2, so are Tf and $\mathsf{M}_{\mathcal{D}}(Tf)$. Therefore, due to Lemmas 6.5 and 6.4 we have $\|\mathsf{M}_{\mathcal{D}}(Tf)\|_p \leq C_{p,n}\|(Tf)_{\mathcal{D}}^{\sharp}\|_p$. We conclude by noticing that $|Tf| \leq \mathsf{M}_{\mathcal{D}}(Tf)$ almost everywhere.

8.2 ▪ Complex methods

8.2.1 ▪ The Riesz–Thorin convexity theorem

Theorem 8.6 (Riesz–Thorin). *Let $(X, \mathcal{A}, \mu)$ and $(Y, \mathcal{B}, \nu)$ be two measure spaces. Let T be a linear map from simple $\mathcal{A}$-measurable functions to $\mathcal{B}$-measurable functions of type (p_0, q_0) with constant M_0 and of type (p_1, q_1) with constant M_1, where p_0, p_1, q_0, and q_1 are in the interval $[1, +\infty]$. Let $\theta \in [0, 1]$ and consider p and q so defined:*

$$\frac{1}{p} = \frac{1-\theta}{p_0} + \frac{\theta}{p_1} \quad and \quad \frac{1}{q} = \frac{1-\theta}{q_0} + \frac{\theta}{q_1}.$$

Then, for all simple $\mathcal{A}$-measurable functions, we have

$$\|Tf\|_{L^q(\nu)} \leq M_0^{1-\theta} M_1^{\theta} \|f\|_{L^p(\mu)}.$$

In particular, T extends as a bounded operator from $L^p(\mu)$ to $L^q(\nu)$.

Proof. We perform the proof in the case where $p \neq +\infty$ and $q \neq 1$. Fix f simple such that $\|f\|_{L^p(\mu)} = 1$. Let g be a simple function such that $\|g\|_{L^{q'}(\nu)} = 1$. We have

$$f = \sum_j a_j \mathbf{1}_{A_j} \quad and \quad g = \sum_k b_k \mathbf{1}_{B_k},$$

where both these sums have a finite numbers of terms, the a_j are all distinct and nonzero, as are the b_k, and the measurable sets A_j are disjoint and have finite measures, and the same for the B_k.

Define two functions α and β of one complex variable:

$$\alpha(z) = \frac{1-z}{p_0} + \frac{z}{p_1} \quad and \quad \beta(z) = \frac{1-z}{q_0'} + \frac{z}{q_1'}.$$

Set

$$f_z = \sum_j \frac{a_j}{|a_j|} |a_j|^{\frac{\alpha(z)}{\alpha(\theta)}} \mathbf{1}_{A_j} \quad and \quad g_z = \sum_k \frac{b_k}{|b_k|} |b_k|^{\frac{\beta(z)}{\beta(\theta)}} \mathbf{1}_{B_k}.$$

Then one can check the following equalities, valid for all real y:

$$|f_{iy}|^{p_0} = |f|^p, \quad |f_{1+iy}|^{p_1} = |f|^p, \quad |g_{iy}|^{q_0'} = |g|^{q'}, \quad and \quad |g_{1+iy}|^{q_1'} = |f|^{q'}.$$

Therefore

$$\|f_{iy}\|_{p_0} = \|f_{1+iy}\|_{p_1} = \|g_{iy}\|_{q_0'} = \|g_{1+iy}\|_{q_1'} = 1. \tag{8.1}$$

Consider the function $\Phi(z) = \int g_z \, T f_z \, d\nu$. As we have

$$\Phi(z) = \sum_{j,k} \frac{a_j}{|a_j|} \, |a_j|^{\frac{\alpha(z)}{\alpha(\theta)}} \frac{b_k}{|b_k|} \, |b_k|^{\frac{\beta(z)}{\beta(\theta)}} \int_{B_k} T \mathbf{1}_{A_j} \, d\mu,$$

the function Φ is entire and bounded on the strip $S = \{z \in \mathbb{C} \ : \ 0 \leq \Re z \leq 1\}$. Moreover, due to (8.1), for all $y \in \mathbb{R}$, we have

$$|\Phi(\mathrm{i}y)| \leq \|T f_{\mathrm{i}y}\|_{q_0} \|g_{\mathrm{i}y}\|_{q_0'} \leq M_0$$

and

$$|\Phi(1 + \mathrm{i}y)| \leq \|T f_{1+\mathrm{i}y}\|_{q_1} \|g_{1+\mathrm{i}y}\|_{q_1'} \leq M_1.$$

Finally, the three-line theorem yields $|\Phi(\theta)| \leq M_0^{1-\theta} M_1^{\theta}$, which means $\left| \int g \, T f \, d\nu \right|$ $\leq M_0^{1-\theta} M_1^{\theta}$. As this holds for any g, we have $\|T f\|_q \leq M_0^{1-\theta} M_1^{\theta}$. The cases left apart are handled in a similar way.

8.2.2 ▪ Two applications

Lemma 8.7 (Young inequality). *Let p, q, and r be three numbers in $[1, +\infty]$ such that $r^{-1} = p^{-1} + q^{-1} - 1$. Then if $f \in L^p$ and $g \in L^q$, we have*

$$\|f * g\|_r \leq \|f\|_p \|g\|_q.$$

Proof. Fix f and consider the operator $T \ : \ g \longmapsto f * g$. Then T is bounded from L^1 to L^p and from $L^{p'}$ to L^∞ with norms bounded from above by 1. The general case comes by applying the Riesz–Thorin theorem.

For an alternate proof of this inequality see Exercise 8.1.

Theorem 8.8 (Hausdorff–Young inequality). *If $1 \leq p \leq 2$ and $f \in L^p$, then $\|\widehat{f}\|_{p'} \leq \|f\|_p$.*

Proof. The proof results from interpolation between the cases $p = 1$ and $p = 2$.

8.2.3 ▪ Complex interpolation with BMO

Let S stand for the closed strip $\{z \in \mathbb{C} \ : \ 0 \leq \Re z \leq 1\}$. We consider a map $z \longmapsto T_z$ from S to the bounded linear operators on $L^2(\mathbb{R}^n)$ with the following properties:

1. For all f and g in $L^2(\mathbb{R}^n)$, $z \mapsto \int g(x) \, T_z f(x) \, dx$ is continuous.

2. For all f and g in $L^2(\mathbb{R}^n)$, $z \mapsto \int g(x) \, T_z f(x) \, dx$ is analytic in the open strip.

This is what is called an *analytic family of operators*.

Theorem 8.9. *We assume moreover that the analytic family T_z fulfills the following requirements:*

1. $M_0 = \sup_{y \in \mathbb{R}} \|T_{iy}\|_{L^2 \to L^2} < \infty$.

2. *For all* $f \in L^2 \cap L^\infty$, $\sup_{y \in \mathbb{R}} \|T_{1+iy}f\|_{\mathrm{BMO}} \leq M_1 \|f\|_\infty < \infty$.

Then, for all $\theta \in \,]0,1[$, *and for all* $f \in L^2 \cap L^p$, *where* $p = \dfrac{2}{1-\theta}$, *we have*

$$\|T_\theta f\|_p \leq C_\theta M_0^{1-\theta} M_1^\theta \|f\|_p,$$

where C_θ depends only on n and θ.

Proof. The number θ is fixed. It is enough to consider simple functions: let us consider a function of the form $f = \sum a_i \mathbf{1}_{A_j}$, where there are finitely many a_j, which are nonzero and all distinct, and such that $\|f\|_p = 1$.

Let m be a positive integer and set

$$T_m^* f(x) = \sup_{Q \in \mathcal{D}_m(x)} \frac{1}{|Q|} \int_Q |T_\theta f(y) - (T_\theta f)_Q| \, \mathrm{d}y, \tag{8.2}$$

where $\mathcal{D}_m(x)$ is the collection of dyadic cubes containing x and whose side has a length between 2^{-m} and 2^m.

The decisive step is to prove the inequality

$$\|T_m^* f\|_p \leq C \, M_0^{1-\theta} M_1^\theta, \tag{8.3}$$

where C depends only on n.

Assuming it has been proven, the proof continues as follows. Since $T_m^* f$ increases towards $(T_\theta f)_{\mathcal{D}}^\sharp$ as m increases to ∞ we get

$$\|(T_\theta f)_{\mathcal{D}}^\sharp\|_p \leq C \, M_0^{1-\theta} M_1^\theta.$$

However, as $T_\theta f$ is in L^2, we get $\|\mathsf{M}_{\mathcal{D}} T_\theta f\|_p \leq C_\theta \, M_0^{1-\theta} M_1^\theta$ (see Corollary 6.6). As $|T_\theta f| \leq \mathsf{M}_{\mathcal{D}} T_\theta f$ almost everywhere, the proof is complete.

Now it is time to prove (8.3). It is clear that the value $T_m^* f(x)$ is attained for some dyadic cube Q_x. If x and x' are in the same dyadic cube of side 2^{-m}, then they are in the same dyadic cube of side 2^k for all $k \geq -m$. This means that $\mathsf{Q}_x = \mathsf{Q}_{x'}$. So, the map $x \mapsto \mathsf{Q}_x$ is measurable. Moreover, for each cube Q_x one can choose a function η_x of modulus 1 such that

$$\frac{1}{|\mathsf{Q}_x|} \int_{\mathsf{Q}_x} |T_\theta f(y) - (T_\theta f)_{\mathsf{Q}_x}| \, \mathrm{d}y = \frac{1}{|\mathsf{Q}_x|} \int_{\mathsf{Q}_x} \big(T_\theta f(y) - (T_\theta f)_{\mathsf{Q}_x}\big) \, \eta_x(y) \, \mathrm{d}y. \tag{8.4}$$

Let U_z ($z \in S$) be the *linear* operator on $L^2(\mathbb{R}^n)$ so defined:

$$U_z h(x) = \frac{1}{|\mathsf{Q}_x|} \int_{\mathsf{Q}_x} \big(T_z h(y) - (T_z h)_{\mathsf{Q}_x}\big) \, \eta_x(y) \, \mathrm{d}y.$$

Proving (8.3) is equivalent to proving that

$$\left| \int U_\theta f(x)\, g(x)\, \mathrm{d}x \right| \leq C\, M_0^{1-\theta} M_1^\theta$$

for any $g = \sum b_k \mathbf{1}_{B_k}$ such that $\|g\|_{p'} = 1$.

Pick such a g and define

$$f_z = \sum \frac{a_j}{|a_j|}\, |a_j|^{\frac{1-z}{1-\theta}}\, \mathbf{1}_{A_j} \quad \text{and} \quad g_z = \sum \frac{a_k}{|b_k|}\, |b_k|^{\frac{1+z}{1+\theta}}\, \mathbf{1}_{B_k}.$$

One has

$$f_\theta = f, \qquad g_\theta = g,$$
$$|f_{\mathrm{i}y}| = |f|^{\frac{p}{2}}, \qquad |f_{1+\mathrm{i}y}| \leq 1,$$
$$|g_{\mathrm{i}y}| = |g|^{\frac{p'}{2}}, \qquad |g_{1+\mathrm{i}y}| \leq |g|^{p'}.$$

Set

$$\Phi(z) = \int U_z f_z(x)\, g_z(x)\, \mathrm{d}x.$$

As $\Phi(z) = \sum \frac{a_j}{|a_j|} \frac{b_k}{|b_k|}\, |a_j|^{\frac{1-z}{1-\theta}} |b_k|^{\frac{1+z}{1+\theta}} \int U_z \mathbf{1}_{A_j}(x)\, \mathbf{1}_{B_k}(x)\, \mathrm{d}x$, one sees that Φ is continuous and bounded on S, and analytic on the open strip. It results from the three-line theorem that

$$|\Phi(\theta)| \leq \left(\sup_{y \in \mathbb{R}} |\Phi(\mathrm{i}y)| \right)^{1-\theta} \left(\sup_{y \in \mathbb{R}} |\Phi(1 + \mathrm{i}y)| \right)^\theta.$$

One has $\|g_{\mathrm{i}y}\|_2^2 = \|g\|_{p'}^{p'} = 1$ and $\|f_{\mathrm{i}y}\|_2^2 = \|f\|_p^p = 1$, so $\|\mathsf{M}_{\mathcal{D}} T_{\mathrm{i}y} f_{\mathrm{i}y}\|_2 \leq C M_0$ for some $C \geq 1$ depending only on n. But $|U_{\mathrm{i}y} f_{\mathrm{i}y}| \leq 2\mathsf{M}_{\mathcal{D}}(T_{\mathrm{i}y} f_{\mathrm{i}y})$. It results that

$$\sup_{y \in \mathbb{R}} |\Phi(\mathrm{i}y)| \leq 2C\, M_0.$$

On the line $\Re z = 1$, one has the following inequalities:

$$|U_{1+\mathrm{i}y} f_{1+\mathrm{i}y}| \leq (T_{1+\mathrm{i}y} f_{1+\mathrm{i}y})_{\mathcal{D}}^\sharp,$$
$$\|f_{1+\mathrm{i}y}\|_\infty \leq 1,$$
$$\| (T_{1+\mathrm{i}y} f_{1+\mathrm{i}y})_{\mathcal{D}}^\sharp \|_\infty \leq M_1,$$
$$\|g_{1+\mathrm{i}y}\|_1 = \|g\|_{p'}^{p'} = 1.$$

So

$$\sup_{y \in \mathbb{R}} |\Phi(1 + \mathrm{i}y)| \leq M_1.$$

This ends the proof.

8.3 ▪ Exercises

Exercise 8.1 (an alternate proof of the Young inequality). *We may suppose that f and g are nonnegative. Write*

$$f(x-y)g(y) = f(x-y)^{1-\alpha}g(y)^{1-\beta}\left(f(x-y)^{\alpha}g(y)^{\beta}\right)$$

and

$$\int f(x-y)g(y)\,\mathrm{d}y =$$
$$\left(\int f(x-y)^{(1-\alpha)u}\,\mathrm{d}y\right)^{1/u}\left(\int g(y)^{(1-\beta)v}\,\mathrm{d}y\right)^{1/v}\left(\int f(x-y)^{\alpha r}g(y)^{\beta r}\,\mathrm{d}y\right)^{1/r},$$

with $\alpha = p/r$, $\beta = q/r$, $u = p/(1-\alpha)$, and $v = q/(1-\beta)$.

Exercise 8.2. *Let μ and ν be two bounded positive measures on $\mathbb{R}^n$.*

1. *Prove that if $f \in L^1(\mu)$, the measure $(f\mu) * \nu$ is absolutely continuous with respect to $\mu * \nu$. So there exists $Tf \in L^1(\mu * \nu)$ such that $(f\mu) * \nu = (Tf)\mu * \nu$.*

2. *Prove that, for all $p \in [1, +\infty]$ and $f \in L^p(\mu)$, we have*

$$\|Tf\|_{L^p(\mu*\nu)} \le \|\nu\|^{\frac{1}{p}}\|f\|_{L^p(\mu)}.$$

3. *Prove that*

$$\|(\check{f} * \nu)\check{}\|_{L^p(\mu)} \le \|\nu\|^{\frac{1}{p}}\|f\|_{L^p(\mu*\nu)}.$$

Exercise 8.3.

1. *We wish to show that, for any $p \in (1,2]$, there exists C_p such that, for all $u \in L^p(\mathbb{R})$, we have*

$$\left(\int_{\mathbb{R}} |\widehat{u}(y)|^p|y|^{p-2}\mathrm{d}y\right)^{\frac{1}{p}} \le C_p\|u\|_{L^p(\mathbb{R})}.$$

   (a) *Observe that this is true for $p = 2$.*

   (b) *We consider the measure $\mathrm{d}\mu(y) = \dfrac{\mathrm{d}y}{y^2}$ and set $Tu(y) = y\,\widehat{u}(y)$.*

   Show that $\mu(\{|Tu| > t\}) \le \dfrac{1}{t}\|u\|_{L^1(\mathbb{R})}$ for all $t > 0$.

   (c) *Conclude.*

2. *Let p be such that $1 < p < 2$. Let p' be its conjugate exponent: $p' = p/(p-1)$. Show that for all $r \in (p, p')$ there exists A_r such that, for any $u \in L^p(\mathbb{R})$, we have*

$$\int_{\mathbb{R}}\left||y|^{\frac{1}{p'}}\,\widehat{u}(y)\right|^r\frac{\mathrm{d}y}{|y|} \le A_r\|u\|_{L^p(\mathbb{R})}^r.$$

Hint: Consider first the cases $r = p$ and $r = p'$.

Appendix A

Background material

A.1 ▪ Vector-valued integrals

Let E be a separable Banach space whose norm is denoted by $|\ |$. Its dual space is denoted by E' and the duality form by $\langle\ ,\ \rangle$.

Let $(X, \mathscr{A}, \mu)$ be a measure space. A function f from X to E is said to be measurable if for any $u \in E'$ the function $\langle f, u \rangle$ is $\mathscr{A}$-measurable. Due to the separability hypothesis, $|f|$ is measurable if f is.

If $\int |f|\,\mathrm{d}\mu$ is finite,

$$u \longmapsto \int \langle f(x), u \rangle\ \mathrm{d}\mu(x)$$

defines a linear form on E' bounded by $\int |f|\,\mathrm{d}\mu$. Therefore there exists a unique element, that will be denoted by $\int f\,\mathrm{d}\mu$, of E'', the bidual of E, such that

$$\int \langle f(x), u \rangle\ \mathrm{d}\mu(x) = \left\langle \int f\,\mathrm{d}\mu, u \right\rangle$$

for all $u \in E'$.

We have $\left| \int f\,\mathrm{d}\mu \right| \leq \int |f|\,\mathrm{d}\mu$. We define the space $L^p(\mu, E)$ to be the space of E-valued measurable functions f such that $\int |f|\,\mathrm{d}\mu$ is finite. We also define as usual the $L^\infty(\mu, E)$ space.

A.2 ▪ Convolution

The convolution of two functions f and g on $\mathbb{R}^n$ is defined as

$$f * g(x) = \int f(x - y)g(y)\,\mathrm{d}y \tag{A.1}$$

when it makes sense.

If $1 \le p, q, r \le +\infty$, $r^{-1} = p^{-1} + q^{-1} - 1$, $f \in L^p$, and $g \in L^q$, then the integral (A.1) is absolutely convergent for almost every x, and one has

$$\|f * g\|_r \le \|f\|_p \|g\|_q.$$

If $p = 1$ and $q = \infty$, $f * g$ is bounded and uniformly continuous. If $1 < p < \infty$ and $q = p'$, $f * g$ is continuous and vanishes at infinity.

A.3 ▪ Polar coordinates

There exists a Borel measure σ_{n-1} on S_{n-1} such that

$$\int_{\mathbb{R}^n} f \, d\mathcal{L}_n = \iint_{(r,x) \in (0+\infty) \times S_{n-1}} f(rx) r^{n-1} dr \, d\sigma_{n-1}(x)$$

if f is measurable and nonnegative, or if f is integrable.

In particular, if f is radial, i.e., $f(x) = \varphi(|x|)$,

$$\int_{\mathbb{R}^n} \varphi(|x|) d\mathcal{L}_n = \mathsf{s}_{n-1} \int_0^{+\infty} r^{n-1} \varphi(r) dr,$$

where $\mathsf{s}_{n-1} = \dfrac{2\pi^{n/2}}{\Gamma(n/2)}$ is the area of the unit sphere in $\mathbb{R}^n$.

See Exercise A.1 for a possible definition of σ_{n-1}.

A.4 ▪ Distribution functions and weak L^p spaces

A.4.1 ▪ Distribution function

Let $(X, \mathscr{A}, \mu)$ be a measurable space. If φ is an $\mathscr{A}$-measurable function and $t > 0$, set $\varrho_\varphi(t) = \mu(|\varphi| > t)$.

It is clear that if φ_j is a nondecreasing sequence of nonnegative measurable functions converging towards φ, then the sequence ϱ_{φ_j} is nondecreasing and converges towards ϱ_φ.

Lemma A.1. *Let $(X, \mathscr{A}, \mu)$ be a measured space and φ a nonnegative $\mathscr{A}$-measurable function. Then, for all $p > 0$, one has*

$$\int_X \varphi^p \, d\mu = p \int_0^{+\infty} t^{p-1} \varrho_\varphi(t) \, dt.$$

Proof. Consider first the case when φ is simple, i.e., $\varphi = \sum_{j=1}^m a_j \mathbf{1}_{A_j}$, where $0 < a_1 < a_2 < \cdots < a_m$. It is convenient to set $a_0 = 0$. Then $\varrho_\varphi(t) = 0$ if $t \ge a_m$, while $\varrho_\varphi(t) = \sum_{k=j}^m \mu(A_k)$ if $a_{j-1} \le t < a_j$.

We have

$$\int_0^\infty pt^{p-1}\varrho_\varphi(t)\,\mathrm{d}t = \sum_{j=1}^m \left(\int_{a_{j-1}}^{a_j} pt^{p-1}\mathrm{d}t\right)\sum_{k=j}^m \mu(A_k)$$

$$= \sum_{k=1}^m \mu(A_k)\sum_{j=1}^k (a_j^p - a_{j-1}^p)$$

$$= \sum_{k=1}^m \mu(A_k)\sum_{j=1}^k a_j^p - \sum_{k=1}^m \mu(A_k)\sum_{j=1}^{k-1} a_j^p$$

$$= \sum_{k=1}^m a_k^p\mu(A_k) = \int \varphi^p\mathrm{d}\mu.$$

To prove the general case, it is enough to observe that any nonnegative measurable function is the pointwise limit of a nondecreasing sequence of simple functions.

Remark A.2. *If we suppose that the measured space $(X,\mathscr{A},\mu)$ is σ-finite, we can give an alternate proof of this lemma. We have*

$$p\int_0^{+\infty} t^{p-1}\mu(\{\varphi > t\})\,\mathrm{d}t = p\int_0^{+\infty} t^{p-1}\left(\int_X \mathbf{1}_{\{\varphi>t\}}(x)\,\mathrm{d}\mu\right)\mathrm{d}t.$$

By using the Fubini theorem we get

$$p\int_0^{+\infty} t^{p-1}\mu(\{\varphi > t\})\,\mathrm{d}t = \int_X \left(\int_0^{\varphi(x)} pt^{p-1}\mathrm{d}t\right)\mathrm{d}\mu(x) = \int_X \varphi(x)^p\mathrm{d}\mu(x).$$

A.4.2 ▪ Weak L^p spaces

Definition A.3. *If $p > 0$, $L_w^p(\mu)$ is the space of measurable functions f such that*

$$\sup_{t>0} t^p\varrho_f(t) < +\infty.$$

The Chebyshev–Markov inequality yields $L^p \subset L_w^p$, indeed $\mu(|f| > t) \leq t^{-p}\|f\|_p^p$. This inclusion is strict; see Exercise A.9.

A.5 ▪ Laplace transform

Let f be a measurable function on $\mathbb{R}^+$. The interior of the set of complex numbers z such that $\mathrm{e}^{-zt}f(t)$ is integrable is—if it is nonempty—the whole plane or a half-plane of the form $\{z : \Re z > p_0\}$. In this case the function

$$F(z) = \int_0^{+\infty} \mathrm{e}^{-tz}f(t)\mathrm{d}t$$

is analytic on the domain $\{z : \Re z > p_0\}$.

The restriction of F to the interval $(p_0, +\infty)$ is called the *Laplace transform* of f.

If $p > p_0$, $F(p + 2i\pi y)$ is the Fourier transform of the function $f(t)\mathrm{e}^{-pt}$. Due to analyticity, if the Laplace transform of f is null, so is the function $F(p + 2i\pi y)$. Since the Fourier transformation is one-to-one, it results that $f = 0$.

We just proved that the Laplace transformation is injective.

A.6 ▪ Khintchine inequalities

Let $(\varepsilon_j)_{j \geq 0}$ be a sequence of independent Bernoulli variables:

$$\mathbb{P}(\varepsilon_j = 1) = \mathbb{P}(\varepsilon_j = -1) = \frac{1}{2}.$$

Such variables can be realized as functions on $[0, 1]$ endowed with the Lebesgue measure: $\varepsilon_j(x) = \mathrm{sgn}\sin 2^j \pi x$. In this case, such functions are called *Rademacher functions*.

Theorem A.4 (Khintchine inequalities). *Let $(a_j)_{j \geq 0}$ be a square summable sequence. Then*

1. *the series $\displaystyle\sum_{j \geq 0} a_j \varepsilon_j$ converges in L^p for all p;*

2. *for all $p > 0$ there exists C_p independent of the a_j such that*

$$\mathbb{E}\left|\sum_{j \geq 0} a_j \varepsilon_j\right|^p \leq C_p \left(\sum_{j \geq 0} |a_j|^2\right)^{\frac{p}{2}};$$

3. *for all $p > 0$ there exists C_p independent of the a_j such that*

$$\left(\sum_{j \geq 0} |a_j|^2\right)^{\frac{p}{2}} \leq C_p \,\mathbb{E}\left|\sum_{j \geq 0} a_j \varepsilon_j\right|^p.$$

Proof. It is enough to consider the case when the coefficients a_j are real.

When $p < 2$, the second assertion follows from the Hölder inequality. Let ξ_j be independent normal variables of expectation 0 and variance 1. One has $\mathbb{E}(\xi_j \mid \mathrm{sgn}\,\xi_j) = \sqrt{\frac{2}{\pi}}\,\mathrm{sgn}\,\xi_j$. So,

$$\mathbb{E}\left(\sum_{j=0}^{m} a_j \xi_j \,\Big|\, \mathrm{sgn}\,\xi_0, \ldots, \mathrm{sgn}\,\xi_m\right) = \frac{2}{\pi}\sum_{j=0}^{m} a_j \,\mathrm{sgn}\,\xi_j.$$

If $p \geq 2$, the Jensen inequality yields

$$\left|\mathbb{E}\left(\sum_{j=0}^{m} a_j \xi_j \,\Big|\, \mathrm{sgn}\,\xi_0, \ldots, \mathrm{sgn}\,\xi_m\right)\right|^p \leq \mathbb{E}\left(\left|\sum_{j=0}^{m} a_j \xi_j\right|^p \,\Big|\, \mathrm{sgn}\,\xi_0, \ldots, \mathrm{sgn}\,\xi_m\right)$$

and

$$\mathbb{E}\left|\frac{2}{\pi}\sum_{j=0}^{m} a_j \,\mathrm{sgn}\,\xi_j\right|^p \leq \mathbb{E}\left|\sum_{j=0}^{m} a_j \xi_j\right|^p.$$

But the sgn ξ_j are independent Bernoulli variables and $\sum_{j=0}^{m} a_j\xi_j$ is a normal variable of expectation 0 and variance $\sum_{j=0}^{m} a_j^2$, so

$$\left(\frac{2}{\pi}\right)^p \mathbb{E} \left| \sum_{j=0}^{m} a_j\varepsilon_j \right|^p \leq 2^{\frac{p+1}{2}} \Gamma\left(\frac{p+1}{2}\right) \left(\sum_{j=0}^{m} a_j^2 \right)^{\frac{p}{2}}.$$

This proves that the sequence $\sum_{j=0}^{m} a_j\varepsilon_j$ is Cauchy and that

$$\mathbb{E} \left| \sum_{j\geq 0} a_j\varepsilon_j \right|^p \leq 2^{\frac{1-p}{2}} \pi^p \Gamma\left(\frac{p+1}{2}\right) \left(\sum_{j\geq 0} a_j^2 \right)^{\frac{p}{2}}.$$

When $p > 2$ the third assertion is a consequence of the Hölder inequality. For $0 < p < 2$, define $q = 4 - p$; so 2 is the midpoint of the interval $[p, q]$. Letting $X = |\sum a_j\varepsilon_j|$, the function $t \mapsto \log \mathbb{E} X^t$ is convex, so we have

$$\left(\mathbb{E} X^2\right)^2 \leq \mathbb{E} X^p \, \mathbb{E} X^q \leq C_q \, \mathbb{E} X^p \, \mathbb{E} X^2,$$

which ends the proof.

A.7 ▪ Exercises

Exercise A.1. *Consider the map $\gamma : x \mapsto x/|x|$ from the unit ball of $\mathbb{R}^n$ minus the origin to the unit sphere S_{n-1}. Let σ_{n-1} be the measure $n\mathcal{L}_n \circ \gamma^{-1}$, where $\mathcal{L}_n$ is the Lebesgue measure on $\mathbb{R}^n$.*

1. *If A is a Borel subset of S_{n-1}, show that the Lebesgue measure of the set $\{tx : x \in A, \ 0 \leq t \leq r\}$ equals $r^n \sigma_{n-1}(A)/n$.*

2. *Show that, for $0 \leq r_1 \leq r_2$, one has*

$$\mathcal{L}_n\{tx : x \in A, \ r_1 \leq t \leq r_2\} = \sigma_{n-1}(A) \int_{r_1}^{r_2} t^{n-1} \, dt.$$

3. *Show that the sets $\{tx : x \in A, \ r_1 \leq t \leq r_2\}$, where $0 < r_1 < r_2$ and A is a Borel set of S_{n-1}, generate the Borel field of $\mathbb{R}^n$.*

4. *Show that the Lebesgue measure on $\mathbb{R}^n$ is the image of the measure $t^{n-1} dt \otimes d\sigma_{n-1}(x)$ (on the space $\mathbb{R}^+ \times S_{n-1}$) under the map $(t, x) \mapsto tx$. Therefore one has*

$$\int_{\mathbb{R}^n} f \, d\mathcal{L}_n = \iint_{\mathbb{R}^+ \times S_{n-1}} f(tx) \, t^{n-1} dt \, d\sigma_{n-1}(x),$$

provided that f is nonnegative or integrable.

Exercise A.2. *Compute v_n, the volume of the unit ball, and s_{n-1} the area of the unit sphere in $\mathbb{R}^n$.*

Hint: *Use the identity*

$$\int_{\mathbb{R}^n} e^{-\pi|x|^2} dx = 1.$$

Exercise A.3. *Let* $\varphi : (0, +\infty) \to (0, +\infty)$ *be nonincreasing such that*

$$\int_{\mathbb{R}^n} \varphi(|x|) \, dx < +\infty.$$

Show that

1. $\displaystyle\int_{\mathbb{R}^n} \varphi(|x|) \, dx = n v_n \int_0^{+\infty} t^{n-1} \varphi(t) \, dt;$

2. $r^n \varphi(r) = o(1) \quad (r \to 0+ \text{ or } r \to +\infty).$

Exercise A.4.   *Prove the integral Minkowski inequality: for* $1 < p < +\infty$ *and* f *nonnegative and measurable,*

$$\left[\int \left(\int f(x,y) \, d\nu(y) \right)^p d\mu(x) \right]^{\frac{1}{p}} \le \int \left(\int f(x,y)^p d\mu(x) \right)^{\frac{1}{p}} d\nu(y).$$

Hint: *Recall that* $\|\varphi\|_p = \sup_{\|\psi\|_{p'} \le 1} |\int \varphi \psi|$ *(where* $\frac{1}{p} + \frac{1}{p'} = 1$*). Take* $g \ge 0$ *such that* $\|g\|_{L^{p'}(\mu)} = 1$ *and consider* $\int g(x) \left(\int f(x,y) \, d\nu(y) \right) d\mu(x).$

Exercise A.5. *Let* $(x,y) \mapsto (K(x,y))$ *be an* L^2-function on $\mathbb{R}^n \times \mathbb{R}^n$. *Show that, for any* $f \in L^2(\mathbb{R}^n)$, *we have*

$$\left[\int_{\mathbb{R}^n} \left| \int_{\mathbb{R}^n} K(x,y) f(y) \, dy \right|^2 dx \right]^{\frac{1}{2}} \le \|K\|_{L^2(\mathbb{R}^n \times \mathbb{R}^n)} \|f\|_{L^2(\mathbb{R}^n)}$$

(use Exercise A.4)

Exercise A.6.

1. *Let* $\alpha > -1$. *Show that there exists* $C_{n,\alpha} > 0$ *such that for every* $x = (x_1, x_2, \dots, x_n) \in \mathbb{R}^n$, *we have*

$$\int_{S_{n-1}} |x \cdot u|^\alpha d\sigma(u) = C_{n,\alpha} |x|^\alpha.$$

2. *Let* $(X_1, \mathcal{A}_1, \mu_1)$ *and* $(X_2, \mathcal{A}_2, \mu_2)$ *be two measure spaces,* $p \ge 1$, *and let* T *be a bounded linear operation from* $L^p(\mu_1)$ *to* $L^p(\mu_2)$ *with norm less than or equal to* M *mapping real functions to real functions.*

(a) *Show that, for any finite sequence of real functions $f_1, f_2, \ldots, f_k$ in $L^p(\mu_1)$, we have*

$$\left\| \left(\sum |Tf_j|^2 \right)^{1/2} \right\|_{L^p(\mu_2)} \leq M \left\| \left(\sum |f_j|^2 \right)^{1/2} \right\|_{L^p(\mu_1)}$$

(b) *Show that, for any sequence of functions $f_1, f_2, \ldots, f_k, \ldots$ in $L^p(\mu_1)$, we have*

$$\left\| \left(\sum |Tf_j|^2 \right)^{1/2} \right\|_{L^p(\mu_2)} \leq M \left\| \left(\sum |f_j|^2 \right)^{1/2} \right\|_{L^p(\mu_1)}.$$

Exercise A.7.

1. *Let K be a function on $\mathbb{R}_+^2$ such that $K(tx, ty) = t^{-1}K(x,y)$. Assume that*
$$A = \int_0^{+\infty} |K(1,y)| \, y^{-\frac{1}{p}} \mathrm{d}y < \infty \text{ for some } p \in [1, +\infty).$$ *Set $Tf(x) =$*
$$\int_0^{+\infty} K(x,y) f(y) \, \mathrm{d}y = \int_0^{+\infty} K(1,y) f(xy) \, \mathrm{d}y.$$ *Prove Schur's inequality:*

$$\int_0^{+\infty} |Tf(x)|^p \mathrm{d}x \leq A^p \int_0^{+\infty} |f(x)|^p \mathrm{d}x.$$

2. *Prove the following Hardy inequalities:*

$$\left(\int_0^{+\infty} \left(\int_0^x f(y) \, \mathrm{d}y \right)^p x^{-(r+1)} \mathrm{d}x \right)^{\frac{1}{p}} \leq \frac{p}{r} \left(\int_0^{+\infty} (yf(y))^p y^{-(r+1)} \mathrm{d}y \right)^{\frac{1}{p}},$$

$$\left(\int_0^{+\infty} \left(\int_x^{+\infty} f(y) \, \mathrm{d}y \right)^p x^{r-1} \mathrm{d}x \right)^{\frac{1}{p}} \leq \frac{p}{r} \left(\int_0^{+\infty} (yf(y))^p y^{r-1} \mathrm{d}y \right)^{\frac{1}{p}},$$

where $f \geq 0$, $p \geq 1$, and $r > 0$.

Exercise A.8. *Let $(X, \mathscr{A}, \mu)$ be a σ-finite measure space, and let f be a nonnegative measurable function on X.*

1. *Given $t > 0$ we set*

$$g(x) = \begin{cases} f(x) & \text{if } f(x) > t, \\ 0 & \text{otherwise} \end{cases} \quad \text{and} \quad h(x) = \begin{cases} 0 & \text{if } f(x) > t, \\ f(x) & \text{otherwise}. \end{cases}$$

Show the following:

$$\mu\{g > u\} = \begin{cases} \mu(\{f > u\}) & \text{if } u > t, \\ \mu(\{f > t\}) & \text{if } u \leq t, \end{cases}$$

$$\mu\{h > u\} = \begin{cases} 0 & \text{if } u > t, \\ \mu(\{f > u\}) - \mu(\{f > t\}) & \text{if } u \leq t. \end{cases}$$

2. *Show that, for $0 < p < +\infty$ and $0 < u < v < +\infty$, we have*

$$\int_{f>u} f^p \mathrm{d}\mu = p \int_u^{+\infty} t^{p-1}\mu(\{f > t\})\,\mathrm{d}t + u^p\mu(\{f > u\}),$$

$$\int_{f\le v} f^p \mathrm{d}\mu = p \int_0^v t^{p-1}\mu(\{f > t\})\,\mathrm{d}t - v^p\mu(\{f > v\}),$$

$$\int_{u<f\le v} f^p \mathrm{d}\mu = p \int_u^v t^{p-1}\mu(\{f > t\})\,\mathrm{d}t - v^p\mu(\{f > v\}) + u^p\mu(\{f > u\}).$$

Exercise A.9. *Let $(\Omega, \mathcal{B}, \mu)$ be a measure space. Show that if, for some $p > 0$, $f \in L^p(\mu)$, one has*

$$t^p\mu(\{|f| > t\}) = o(1) \ (t \to 0 \text{ or } t \to +\infty).$$

Exercise A.10. *Let $(\Omega, \mathcal{B}, \mu)$ be a σ-finite measure space and let $\varphi : [0, +\infty) \to [0, +\infty)$ be a nondecreasing function such that $\varphi(0) = 0$. Show that, for any measurable function f on Ω, one has*

$$\int_\Omega \varphi(|f|)\,\mathrm{d}\mu = \int_0^{+\infty} \mu(\{|f| > t\})\,\mathrm{d}\varphi(t).$$

Exercise A.11. *Let $(\Omega, \mathcal{B}, \mu)$ be a probability space and let $p_0 > 0$. Show that*

$$L^{p_0}(\mu) \subset L_w^{p_0}(\mu) \subset \bigcap_{0<p<p_0} L^p(\mu).$$

Exercise A.12. *Let $(\Omega, \mathcal{B}, \mu)$ be a measure space, and let $0 < p_0 < p_1$. Show that, for all $p \in (p_0, p_1)$, one has*

$$L_w^{p_0}(\mu) \cap L_w^{p_1}(\mu) \subset L^p(\mu).$$

Exercise A.13. *Let $p > 0$. If $f \in L_w^p(\mu)$, set*

$$a(f) = \sup_{t>0} \left(t^p\mu(\{|f| > t\})\right)^{\frac{1}{p+1}}.$$

1. *For all $f, g \in L_w^p(\mu)$, show that*

$$a(f + g) \le a(f) + a(g)$$

   and that $\mathbf{d}(f, g) = a(f - g)$ defines a distance on $L_w^p(\mu)$.

2. *Show that the space $L_w^p(\mu)$ endowed with the distance $\mathbf{d}$ is complete.*

Exercise A.14. *Let* $\alpha > 0$. *Consider the Hilbert space*

$$\mathcal{H}_\alpha = \left\{ f : \int_0^{+\infty} |f(r)|^2 r^{\alpha-1} dr < +\infty \right\}.$$

Show that the functions $r \mapsto e^{-t^2}$ *($t > 0$) generate* $\mathcal{H}_\alpha$.

Hint. *Use the injectivity of the Laplace transformation to show that a function orthogonal to all* e^{-tr^2} *is null.*

Appendix B

Notation and conventions

B.1 ▪ Glossary of notation and symbols

$\lvert x \rvert$	the Euclidean length of $x \in \mathbb{R}^n$
$\lvert E \rvert$	the Lebesgue measure of E
$x \cdot y$	the Euclidean scalar product
${}^c E$	the complement of the set E
$\widehat{f}$ or $\hat{f}$	the Fourier transform of f: $\widehat{f}(\xi) = \displaystyle\int \mathrm{e}^{-2\mathrm{i}\pi\xi\cdot x} f(x)\,\mathrm{d}x$
$\check{f}$	$\check{f}(x) = f(-x)$
$\tau_a f$	$\tau_a f(x) = f(x - a)$
∇f	the gradient of f
$\lVert f \rVert_{L^p(\mu)}$ or simply $\lVert f \rVert_p$	the norm of f in L^p
$\lVert T \rVert_{p \to q}$	the norm of T as an operator from L^p to L^q
$\lVert\ \rVert_*$	the BMO norm
p'	the exponent conjugate to p, $1 \le p \le +\infty$, $p^{-1} + p'^{-1} = 1$
$\{f \in E\}$	the set $\{x \ : \ f(x) \in E\}$
$\mathrm{B}\,(x, r)$	the Euclidean open ball of radius r centered at x
$\mathrm{BMO}(\mathbb{R}^n)$	the space of functions of bounded mean oscillation
$\mathrm{BMO}_\mathcal{D}$	a dyadic BMO space
$\mathscr{D}(\mathbb{R}^n)$ or simply $\mathscr{D}$	the space of C^∞ functions with compact support
$\mathcal{D}(\mathrm{Q})$	the set of dyadic cubes subordinate to Q
$\mathrm{i} = \sqrt{-1}$	
$\mathscr{K}(\mathbb{R}^n)$	the space of continuous functions with compact support
$L^p(\mu)$ or simply L^p	the space of pth power integrable functions with respect to μ
$\ell^p(I)$ or simply ℓ^p and occasionally $L^p(I)$	the set of $\{a_i\}_{i \in I}$ endowed with the norm $\lVert a \rVert_p = \left(\sum_{i \in I} \lvert a_i \rvert^p\right)^{1/p}$
$\mathcal{L}_n$	the Lebesgue measure on $\mathbb{R}^n$ ($\mathcal{L}$ if no confusion occurs)
$L^p(\mathbb{R}^n)$	the L^p space of $\mathcal{L}_n$
L^p_{loc}	the space of functions f such that $\lvert f \rvert^p$ is locally integrable
L^p_{w}	The space weak L^p

109

M	the Hardy–Littlewood maximal operator
$M_{\mathcal{D}}$	any dyadic maximal operator
p'	the exponent conjugate to p, $1 \le p \le +\infty$, $p^{-1} + p'^{-1} = 1$
$Q(a, \ell)$	the cube of side ℓ whose lower left corner is a, i.e.,

$$Q(a, \ell) = \prod_{j=1}^{n} [a_j, a_j + \ell)$$

S_{n-1}	the unit sphere in $\mathbb{R}^n$
s_{n-1}	the area of the unit sphere S_{n-1} ($s_{n-1} = n v_n$)
sgn	the signum function $\operatorname{sgn} x = \begin{cases} +1 & \text{if } x > 0, \\ 0 & \text{if } x = 0, \\ -1 & \text{if } x < 0 \end{cases}$
$\mathscr{S}(\mathbb{R}^n)$ or simply $\mathscr{S}$	The Schwartz space of C^∞ functions whose all the derivatives are rapidly decreasing
σ_{n-1} or simply σ	the area measure on S_{n-1}
v_n	the volume of the unit ball in $\mathbb{R}^n$

B.2 ▪ Conventions

In this book, homogeneous means positively homogeneous: if f is a function on $\mathbb{R}^n \setminus \{0\}$ it is said to be positively homogeneous of degree k if, for all $t > 0$ and all x, one has $f(tx) = t^k f(x)$. This property is phrased as "f is k-homogeneous" or "f is homogeneous of degree k."

Also, we drop as many parentheses or braces as possible without altering the clarity. For instance, we write

$$\iint_{x^2 + y^2 < 1} f(x, y) \, \mathrm{d}x \, \mathrm{d}y \quad \text{instead of} \quad \iint_{\{(x,y) \,:\, x^2 + y^2 < 1\}} f(x, y) \, \mathrm{d}x \, \mathrm{d}y,$$

and if μ is a measure, f is a measurable function, and E is a measurable set, we write

$$\mu(f \in E) \quad \text{instead of} \quad \mu(\{f \in E\}).$$

Postface

The presentation of the material presented in this book, which now has been common knowledge for forty years, is, as said in the preface, intentionally very short and does not aim at giving a panorama of the state of the art by the end of the 1960s. On the contrary, this is a selection. Nevertheless, the reader might be interested in the historical genesis of these theories or could be in search of a more complete exposition as well as more recent developments. So we provide a very succinct list of books in which such information can be found.

The treatment of 1-dimensional singular integrals, i.e., the Hilbert transform can be found in [10, 4]. It heavily relies on complex analysis. The 1-dimensional Littlewood–Paley theory and Marcinkiewicz multiplier theorem are treated in [10].

The book by E.M. Stein [6] is the first one which gives a comprehensive treatment of singular integrals. This book, extremely rich, has had a profound influence. The book by E.M. Stein and G. Weiss [8] gives another focus on the subject. Both books deal with Euclidean spaces, but in [1] R.R. Coifman and G. Weiss consider singular integrals on some manifolds. Twenty years later E.M. Stein published a new book [7], a summary of the current knowledge, with a very large scope. More recently, L. Grafakos wrote a new summary in two volumes [2, 3] on harmonic analysis, which is now an unavoidable reference. All these books contain exhaustive bibliographies accounting for the genesis of these theories.

The book by W. Rudin [5] deals with analysis on locally compact abelian groups. The one by M.H. Taibleson [9] has a more specialized scope, namely, Fourier analysis on local fields. Both can be useful when studying Chapter 7 of this book.

Bibliography

[1] R.R. Coifman and G. Weiss: Analyse Harmonique Non-commutative sur Certains Espaces Homogènes, Etude de Certaines Intégrales Singulières. Lecture Notes in Math. 242, Springer, 1971. ISBN: 3-540-05703-X.

[2] L. Grafakos: Classical Fourier Analysis, Third Edition. Graduate Texts in Math. 249, Springer, New York, 2014. ISBN: 978-1-4939-1193-6.

[3] L. Grafakos: Modern Fourier Analysis, Third Edition. Graduate Texts in Math. 250, Springer, New York, 2014. ISBN: 978-0-387-09433-5.

[4] Y. Katznelson: An Introduction to Harmonic Analysis, Cambridge University Press, 1968, 2004. ISBN: 0-521-54359-2.

[5] W. Rudin: Fourier Analysis on Groups. Interscience Tracts in Pure and Appl. Math., Interscience Publisher, 1962.

[6] E.M. Stein: Singular Integrals and Differentiability Properties of Functions. Princeton Mathematical Series, 1971. ISBN: 9780691080796.

[7] E.M. Stein: Harmonic Analysis, Real-Variable Methods, Orthogonality, and Oscillatory Integral. Princeton Math. Ser. 43, 1993. ISBN: 9780691032160.

[8] E.M Stein and G. Weiss: Introduction to Fourier Analysis on Euclidean Spaces. Princeton University Press, 1971. ISBN: 978-0691080789.

[9] M.H. Taibleson: Fourier Analysis on Local Fields. Princeton University Press and University of Tokyo Press, 1975. ISBN: 9780691618128.

[10] A. Zygmund: Trigonometric Series I & II. Cambridge Mathematical Library, Cambridge University Press, 1959. ISBN: 9780521358859.

Index